Graphene Nanomaterials

Graphene Nanomaterials

Kal R. Sharma

MP MOMENTUM PRESS

MOMENTUM PRESS, LLC, NEW YORK

First published by Momentum Press®, LLC
222 East 46th Street, New York, NY 10017
www.momentumpress.net

ISBN-13: 978-1-60650-476-5 (hardback, case bound)
ISBN-13: 978-1-60650-477-2 (e-book)

Momentum Press Nanomaterials Collection

Collection ISSN: Forthcoming (print)
Collection ISSN: Forthcoming (electronic)

DOI: 10.5643/9781606504772

Cover design by Jonathan Pennell
Interior design by Exeter Premedia Services Private Ltd.,
Chennai, India

10 9 8 7 6 5 4 3 2 1

Printed in the United States of America

The book is dedicated to R. Hari Subrahmanyan Sharma (alias Ramki-shan), my eldest son, who turns thirteen on August 13th 2014, with love.

Abstract

Graphene Nanomaterials is expected to fill a void in knowledge among practitioners generated by the discovery of graphene as a distinct allotrope of carbon (2010 Nobel Prize in Physics) with the potential to affect further increases in speed of microprocessors beyond 30 peta hertz. It has other interesting performance properties. Identified in 2004, currently the number of patents in graphene is 7,351 and the number is rising rapidly. This book provides information on the synthesis, characterization, application development, scale-up, stability analysis using a pencil and paper, and structure-property relations. With less than 24,000 atoms/25 nm, the nanosheet form is metastable. Thirty-nine different nanostructuring methods were reviewed in an earlier book including epitaxy, lithography, deposition, exfoliation, etc. With the thickness of only a few atomic layers, graphene has superior field emitter properties, is 100 times stronger than steel, flexible as rubber, tougher than diamond, and is 13 times more conductive than copper. Electron mobility in graphene has been found to be 200,000 $cm^2V^{-1}s^{-1}$.

Different methods of fabrication of graphenes are elaborated and the cost of production can be optimized by consideration of total cost as a sum of capital cost and operating costs. High temperatures needed in the reactor may increase the operating costs. The different processes to make graphene that are discussed in this book are roll-to-roll transfer process, low pressure chemical vapor deposition, atmospheric plug flow reactor, APFR performance analysis, dispersion using NMP, exfoliation from carbonizing catalyst, ion implantation and layer thickness control, chemical method, large area synthesis by phase separation, unzipping CNTs and chemical thermal method, nanoribbon alternation, electrophoretic deposition and reduction, flash cooling, gas intercalation and exfoliation, graphene shell formation, and coal tar pitch as source. The diffusion time calculations in intercalation and exfoliation processes are shown for both Fick and hyperbolic diffusion models. The methods of handling surface reactions, sublimation transport, auto-catalysis, and layer transfer are considered. The characterization methods discussed are Raman Spectroscopy, TEM, SEM, SPM, HeIM, EIS, and morphology analysis.

The cost to prepare a single layer of graphene needs to be reduced. Currently, the cost of production of graphenes is high. The cost depends on the substrate used. A 50 × 50 mm monolayer thin film of graphene from Graphene-Square is $250 for copper substrates and $808 for PET substrates. Graphene nanoplatelets (5 to 8 nm thick) are sold at $218 to $240 per kg. Graphene sheets may have an area with length greater than or equal to 1 mm and fall in a range of 1 mm to 1,000 mm along the transverse and longitudinal directions. The electrical, optical, thermal, magnetic, chemical, quantum, and other properties of graphenes are discussed in detail.

Application development of graphene nanomaterials is examined for the following; Supercapacitor, Barristor, Microprocessor Speed Increases Beyond 30 peta Hertz, Pollutant Capture, Heat Conduction, High Capacity Electrodes, Solar Cells – AR Coating, Carbon Composites, Panel in Wireless Telephones, Thermal Managements, Medical Applications, Genome Analysis using Graphene, Drug Delivery, Piezoelectric Sensors, Desalination, Molecular Sieve, Antibacterial, Quantum Dots, and Transistors. The graphene market is expected to grow to a size of $126 million by the year 2020. By the year 2015, the nanotechnology market is expected to be 3 trillion. Triple-junction solar cells with a light-power conversion efficiency of 41% have been reported by Siemens and Semprius Inc.

2D nanosheets cannot be generated without an epitaxial substrate, which can be used to provide atomic bonding in the third dimension (Landau-Pearls argument). Different stability considerations are discussed in detail, which includes, Free Energy Considerations, Enthalpy Considerations, Epitaxial Stability, Landau-Pearls Stability, Euler Stability, Kekule Structure, Mackay's Radius of Gyration, "Negative" and "Positive" Curvature of Sheets, Puckering and Wrinkling, Island Formation, Metastability, Interface Formation, and Surface Tension.

Keywords

single-layer graphenes, barristor, ultracapacitor, carbon allotrope, thinnest material, deposition, milling, scotch tape, honey comb structure, 2D lattice, unscrolled CNT, industrial electronics, nanomaterials, transparent electrodes and other applications, cost of production, roll-to-roll transfer and other fabrication processes, APFR, diffusion times, Raman spectroscopy, TEM, HeIM and other characterization methods, hexagonal anion rings, magnetic, surface, electrical, and mechanical properties, quantum hall effect, electrorheological properties, catalysts, thermodynamic stability-free energy of reaction, scroll stability, surface reactivity, interfacial stability, edge stability, metastability, defects

Contents

Preface

Graphene has been discovered as a distinct allotrope of carbon (2010 Physics Nobel Prize). It has a two-dimensional unique hexagonal lattice structure made of planar sheets of sp^2-hybridized carbon atoms different from the Bravais lattices known in materials science. *Graphene Nanomaterials* describes the discovery, prospects, characterization methods, applications, stability considerations, fabrication methods, and properties of graphene. It possesses interesting electronic, optical, mechanical, and thermal properties. A number of interesting applications are expected for single-layer graphene in the areas of computing, energy, and medicine. Interest in development of graphene is increasing worldwide including investments in the European Union, Russia, Korea, and the United States (National Nanotechnology Initiative). It has the potential to effect further increases in microprocessor speeds beyond 30 pHz. It has other interesting performance properties. Identified in 2004, the number of patents on graphene is 7,351 and rising rapidly. Less than 24,000 atoms/25-nm nanosheet form is metastable. Thirty-nine different nanostructuring methods were reviewed in an earlier book including epitaxy, lithography, deposition, exfoliation, and so forth. One or more atom layers in thickness, graphene has superior field emitter properties, is 100 times stronger than steel, flexible as rubber, tougher than diamond, and is 13 times more conductive than copper. Electron mobility in graphene has been found to be 200,000 $cm^2\ V^{-1}\ s^{-1}$.

According to a recent Lux report, the projected market value of graphene by 2018 is $180 million. According to the British Broadcasting Corporation (BBC), by 2020, the market value of graphene will be $675 million. The Lux report did not include an economically scalable model of fabrication of graphene in their estimates. A number of scalable methods to make graphene are discussed in Chapter 5. The cost of production of graphene is expected to come down as the technologists move past the learning curve. It costs $60 per square inch of graphene on a copper substrate. Expectations are high for the costs to come down to $1 per square inch for industrial electronic applications and 10 cents per square inch for use in touch-screen displays. Different methods of fabrication of graphenes are ellaborated. In processes where operating costs are high, an optimal cost solution may exist. The different processes to make graphene that are discussed are roll-to-roll transfer process in an atmospheric plug flow reactor, dispersion using N-methyl-2-pyrrolidone, exfoliation from a carbonizing catalyst, ion implantation and layer thickness control, chemical method, large-area synthesis by phase separation, unzipping carbon nanotubes, and chemical–thermal method, nanoribbon alternation, electrophoretic deposition and reduction, flash cooling, gas intercalation and exfoliation, graphene shell formation, and coal tar pitch as the source. The diffusion time calculations in intercalation and exfoliation processes are shown for both Fick and hyperbolic diffusion models. It is shown how to handle surface reactions, sublimation transport, autocatalysis, and layer transfer.

The characterization methods described are Raman spectroscopy, helium ion microscopy, small-angle X-ray scattering, transmission electron microscopy, surface electron microscopy, scanning probe microscopy, microwave spectroscopy, Auger electron microscopy, X-ray diffraction, and others. Application development of graphene nanomaterials is discussed in detail for the following: supercapacitors, desalination, light-emitting diodes, thermal management, transparent electrodes, solar cells, batteries, anticorrosion coating, bionic materials, electromagnetic shielding, oil spills, superconductors, rapid DNA sequencing, magnetic sensors, nanorobots, and nanoscale thermometers. Chemical modification and Rusnano initiative are also discussed. Two-dimensional nanosheets cannot be generated without an epitaxial substrate, which can be used to provide atomic bonding in the third dimension (Landau–Peierls argument). Different stability considerations are discussed in detail that include thermodynamic stability—free energy of reaction, scroll stability, surface reactivity, interfacial stability, edge stability, metastability, and defects.

The magnetic, surface, electrical, and mechanical properties of graphenes are discussed. Quantum hall effect, electrorheological properties, hexagonal onion rings, and role in catalysts of graphenes are also examined.

CHAPTER 1

Discovery and Prospects

Chapter Objectives

- What is graphene?
- Carbon allotropes
- Graphene in news
- Market potential
- Applications
- News
- Investments in China and Russia

Carbon Allotropes

Graphene[1] is a distinct allotrope of carbon. The Nobel Prize in Physics was awarded to Prof. A. K. Geim and K. Novoselov in 2010 for their research on graphenes. Other allotropes of carbon that have received considerable attention and are of interest are as follows: (1) graphite;[2] (2) diamond;[3] (3) fullerene,[4] C_{60}; and (4) carbon nanotubes (CNTs).[5] Graphite was discovered in 1564. The energy that binds the interlayer graphene sheets is around 5.9 kJ mol^{-1}. Graphene can be obtained by exfoliating layers from the bonded sheets down to one layer. Carbon nano-foam[6] has been identified as another form of carbon. The Nobel Prize in Chemistry was awarded to Sir Harry Kroto, Richard Smalley, and Robert Curl in 1996 for their research on fullerenes. Other forms of carbon mentioned in the literature, such as lonsdaleite found in meteorites, C_{540}, and fullerene nanobuds, need better characterization and further understanding before being discussed. Vitreous carbon is a non-graphitizing carbon and can be used to make high-temperature crucibles. Substantial portions of bituminous coals[7] are made of fixed carbon (FC). Dehydrogenation of the FC residue can lead to another form of carbon. Charcoal, soot, amorphous carbon, and coke formed during catalytic reactions are also examples of sources containing carbon in predominant portions.

Fullerenes are made of sheets of pentagons and hexagons of carbon and obey the Euler stability criterion. The sheets have positive curvature and form a soccer-ball structure. In carbon nanofoams, a heptagonal structure with "negative curvature" is seen.

Graphene in News

The two-dimensional lattice structure of graphene is unique and different from the 14 three-dimensional Bravais lattices known in materials science. It can be considered a hexagonal two-dimensional lattice. Graphene is found to be superstrong. It comprises one layer of honeycomb-structured atoms. Some consider that the discovery of graphene is the next big development since the invention of James Watt's steam engine at the onset of industrial revolution and John Bardeen's invention of transistor and superconductor in the 20th century. A number of jobs can be generated by the development of graphene. Graphene is the subject of investigation in every top-notch university around the world. Multinational conglomerates are developing vision and mission plans that are geared toward market capture of graphene products.

The American Chemical Society has called graphene a "wonder material." It is a two-dimensional crystal. It possesses interesting electronic, optical, mechanical, and thermal properties. It has been called as "miracle material" because it was found to have high strength and self-healing properties. Graphene has been found to be the strongest material ever produced, surpassing another carbon allotrope, diamond; it is a conductor of electricity and has self-cooling properties.

Expectations are high for graphene to be used in applications such as solar cells,[8] cameras, transistors, computers, and touch-screen devices. With some modification, it can replace silicon in barristers and transistors. Graphene is highly reactive. A separate chapter is devoted to stability considerations from different perspectives.

The European Union is investing €1 billion as funding for 10 years in order to explore commercial applications of graphene. The Russian initiative plans to spend $8.55 billion in order to create a nanotech industry by the year 2015. More about Rusnano is discussed in the section "Investments on Nanotechnology in Russia." According to Grafoid, in Ottawa, Canada, there are 7,500 patents related to graphene. China currently has a plurality of ownership of intellectual properties in this area followed by the United States and South Korea. Cumulative patents on graphene over the past 5 years have quintupled.

The Chinese Academy of Sciences is working on China occupying a dominant position in use of graphene. Graphene industrial zones are established on a regional basis throughout China. The importance of graphene is recognized by the United States, and a number of R&D projects are funded at the university and corporate levels. Three top graphene developers are International Business Machines Corp. (IBM), Xerox Corp., and Samsung.

South Korea has apportioned a $350 million initial amount for graphene development funding. Graphene can be used as an export-rich catalyst and can contribute to the economic, industrial, and manufacturing expansion. Canada has started a public–private consortium to foster graphene-printed inks. Canada is developing supercapacitors from graphenes. These can be used in renewable energy projects.

BASF and Max Planck Institute for Polymer Research have set up a joint research laboratory for graphene. They have invested €10 million in a carbon innovation center in 2012. The team of investigators is planning to conduct research on carbon-based materials for use in energy storage systems and electronic applications. The 200-m² laboratory is used by an international interdisciplinary team of chemists, physicists, and material scientists. They are studying the potential of graphene and other innovative carbon-based materials. They are exploring applications in batteries, catalysts, and graphene-based composites.

Massachusetts Institute of Technology (MIT), Cambridge, MA, and Microsystems Technology Laboratories announced in 2011 the creation of the Center for Graphene Devices and Systems. Their mission is to advance the science and engineering of graphene-based technologies. They are exploring applications in energy generation, smart fabrics and materials, radio-frequency communications, and sensing.

IBM in 2011 announced that it has designed high-speed circuits from graphene. One such circuit was reported in the journal *Science* that can be used as a frequency mixer. The circuit is built on a wafer of silicon. Signals with one frequency can be shifted to another frequency using the circuit. Frequency mixing was demonstrated up to speeds of 10 GHz. Several layers of graphene were deposited on a silicon wafer. IBM has used inductors as components in the circuit. Over $1.5 billion has been invested in South Korea and European Union on the next-generation display material. Graphene is explored as a substitute for materials such as gallium arsenide used in high-frequency military communication equipment. Graphene can be used in wireless applications in transfer of high-resolution video and data. The life span of organic light-emitting diodes (OLEDs) can be expanded by use of graphene in display devices. Graphene films introduced by the University of Texas at Austin were scaled up in South Korea at Sungkyunkwan University in order to make full-screen displays. A silicon carbide wafer was heated to 1,300°C at IBM, allowing the silicon atoms on the surface to evaporate and the remaining carbon atoms to form the hexagonal graphene sheet. The cost of silicon carbide is high. IBM is exploring ways to reduce the production cost.[9]

Honeycomb Lattice Structure

The honeycomb lattice structure has been confirmed for graphenes. A transmission electron micrograph (TEM) of Graphene X™ is shown in Figure 1.1. The hexagonal arrangement of atoms is striking and hard to miss! The hexagonal sheets of atoms are planar. The electrons in the sp^2-hybridized orbitals get delocalized. The delocalization of electrons in a monolayer graphene sheet is shown in Figure 1.2. Electrons from the top row of hexagonal rings flow to the second layer due to delocalization. This phenomenon was first observed in benzene and was called the Kekulé structure. The discovery was made in the year 1865. Alternating single and double bonds can be seen in the hexagonal rings. In order to maintain the octet configuration, some hexagonal rings have three alternate single and three double bonds, some have two double bonds, and the rest have single bonds in an alternating manner. All carbons have an octet configuration. Further reduction of unsaturation may be attained by polymerization by scientists. Electrons can flow readily without any obstacle. This can lead to interesting electrical properties of the material. Different morphologies of graphene are possible in addition to the sheet form such as chiral, armchair, and puckered.

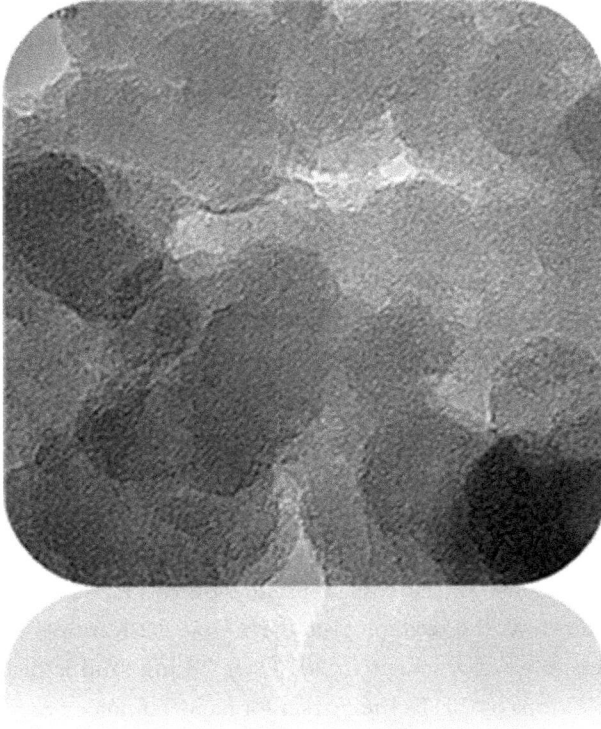

Figure 1.1 A transmission electron micrograph with 5-nm resolu-
tion of nanosheets (technical data sheet of Graphene X^{TM}—Grade 1,
Graphene Technologies, www.graphenetechnologies.com)

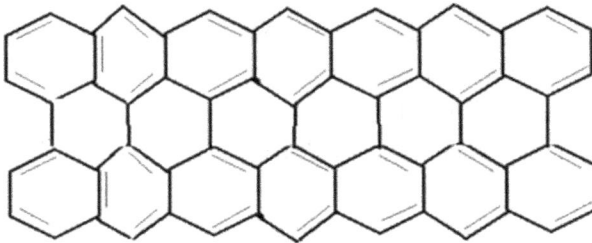

Figure 1.2 Delocalized p-orbital electrons in a graphene monolayer

A technologist from Columbia University calculated that it would
take an elephant balanced on a pencil to puncture a graphene sheet
of the thickness of the saran wrap. A workshop was conducted by the

National Science Foundation and Air Force Office of Sponsored Research in Arlington, VA, in 2012 in order to bring together leading scientists to discuss further impacts of two-dimensional graphene atomic layers and devices and the technological and scientific implications.

Market

According to a recent Lux Research report, the market value of graphene is $10 million. According to the British Broadcasting Corporation (BBC), the graphene market is expected to grow to $67 million by the year 2015 and to $675 million by the year 2020. According to the Lux Research report, the projected market value of graphene by 2018 is $180 million. The reason for the lower estimate of Lux is that they did not include an economically scalable model of graphene fabrication in their estimates. A number of scalable fabrication methods of graphene are discussed in Chapter 5. The cost of production of graphene is expected to come down as the technologists move past the learning curve. The production value of graphene in 2010 was 28 tons and is projected to grow to 573 tons by 2017. The market for nanotechnology products will reach $3 trillion by the year 2015. Leading players such as IBM, Dow Corning, Intel, and Boeing are planning to introduce terahertz (THz) graphene processors. This can lead to graphene market size reaching the billion level.

The potential for applications of graphene is tremendous. The applications are discussed in detail in Chapter 3. Graphenes can be used in computing, energy, and medicine. These include a wide range: high-capacity electrodes, antireflection coatings in solar cells, carbon composites for light-weight BMWs, panel displays in wireless telephones and laptops, and thermal management. Graphene may be used in cutting-edge cancer treatments and in feather-weight, high-definition (HD) televisions. Monolayer graphene is gapless. In order to have a tunable bandgap, another layer of graphene is needed. Graphene can be used in the study of RNA and DNA sequences and speedy genome completions. Graphene-based transistors are being developed. Novel magnetologic gates (MLGs) are designed that can obviate the von Neumann bottleneck. Bilayer graphene can be used to provide tunable bandgaps. Graphene can be used in supercapacitors, desalination devices, light-emitting diodes, metal matrix composites, heat sinks, transparent electrodes, solar cells, batteries, anticorrosion coatings, bionic materials, infrared transparent materials, electromagnetic shielding coatings in naval applications, oil spill abatement devices, graphene nanocomposites, biological sensors with some modification, superconductors, magnetic sensors, superfast lasers, drug delivery systems, cell imaging systems, and other biomedical applications.

Graphene is projected to be used in conductive, programmable construction wall panels. They may be incorporated into holographic solar panels that can be used to follow the sun's rays and used in organic electronics, organic electronics and lighting, transparent conductive films, touch screens, and so forth. An estimated 50 billion people worldwide use portable electronic devices such as smartphones, tablets, and laptops. These devices will need rechargeable batteries. Graphene can be used to make better batteries. Graphenes can be used as electrodes in fuel cells. The market for fuel cells is worth about $2 billion.

Graphene is a two-dimensional natural substance and can be obtained from graphite flakes. Graphene is one-atom thick. It is transparent, flexible, and stretchable and is found to be 200 times stronger than steel. Maintaining its tensile strength of 1.5 million psi, graphene can be stretched to the size of a football field. Graphene has superior electrical conduction properties compared with copper and silver. Graphene can be used to annihilate bacteria and to effect a sterile environment for curing burn victims by enabling growth of skin cells.

International standards are yet to be established for graphene. The melting point of graphene is 3,000°C. This makes it suitable for use as a material of construction in rocket nozzles and heat shields. This can better the preceding material used, that is, graphite. Graphene coating on wires can be used to decrease the electrical transmission losses.

Graphene can be formed by unscrolling of CNTs. It looks like a molecular chicken wire. Graphene with its one-atom thickness can be used at room temperature to study quantum phenomena. One hundred papers were presented at an American Physical Society meeting at Denver, CO, recently.[10] Students in the materials science class learn about graphite comprising several layers of planar sheets of carbon held together by van der Waals forces. The black trail produced using a pencil with graphite lead is due to rubbed-off graphite flakes. Layers of carbon atoms stacked on top of each other as a deck of cards can be hypothesized from observation of the black trail. Prof. R. Ruoff who is currently at the University of Texas at Austin has reported that when tiny pillars of graphite are rubbed against a silicon wafer surface, they spread out like a deck of cards. Single-layer graphene can be produced using this technique. Prof. P. Kim at the Physics Department of Columbia University made a nanopencil by attaching a graphite crystal to the tip of an atomic force microscope (AFM) and dragging it along the surface. Flakes cleaved from graphite. The thickness of the flakes was 5 billionths of a meter and comprised of maybe 10 layers of atoms. Using a simple light microscope, researchers can gauge the number of layers of graphene in the sample, that is, whether they are 10-layer thick, 30- to 40-layer thick, single-layer thick,[11] and so forth. The 100-layer-thick sample was yellow, 30- to 40-layer-thick sample was blue, 10-layer-thick sample was pink, and single-layer-thick sample was pink. Prof. A. K. Geim, a Nobel Prize recipient (2010), observed color change of the silicon oxide layer atop the wafer with single-layer graphene. Dr. Bennakker at Leiden University proposed "valleytronics," where delocalized electrons in graphene are tapped into.

The discovery of graphene has sparked a global scientific gold rush. Companies and universities have joined the race to understand and obtain intellectual property rights for manipulation and application of the thin graphene. Electronic newspapers that can be folded into a pocket and flexible phones can be made from graphenes. A graphene-induced tennis racket has been introduced in the market. A team of 40 scientists is working at the Cambridge University, Cambridge, United Kingdom, on graphenes. The hunt is on for finding the value of the graphene material.

Work is in progress to reduce the cost of mass production of graphene. It costs $60 to make a square inch of graphene on a copper substrate. The cost would have to come down to $1 per square inch for it to be used in industrial electronic applications and 10 cents per square inch for use in touch-screen displays. It has to be modified into a usable form in industrial electronic applications. Apple has patented an application of graphene as a heat dissipator in mobile devices. Graphene can be used in heating circuits for deicing airplane wings. A patent for seawater desalination using micropores in graphene membranes was filed by Lockheed Martin.[12]

The circuitry laid down using graphene ink has ignited the interest of investors. Graphene ink has been used to print circuitry that can go into upholstery that can be used to heat car seats according to a BASF patent. Graphene ink can be used in antitheft devices.

There is increased interest in graphene as the base material for nano-electromechanical systems (NEMS). Lightness and stiffness are important requirements of NEMS devices. Low inertial masses, ultrahigh frequencies that can be achieved, and low-resistance contacts make graphene an attractive choice for NEMS circuits. Young's modulus comparable to that of graphene at room temperature is seen in nanometer-thick polycrystalline NEMS. Quality factors of ~4,000 are seen.[13] A. K. Geim (Nobel Prize, 2010) has called graphene a "sleeping beauty."

Short-term gain from development of graphenes is achieved in the area of high-speed electronics and flexible circuitry. A researcher from Rice University has pointed out that graphene can be made from different sources such as grass on a lawn in the backyard, Girl Scout Cookies, and cockroach legs.

Graphene composites can be used in natural-gas tanks and in exterior trim of automobiles. IBM researchers found that the electrons travel too fast to be used as a switch and hence conversion of "ones" and "zeros" in binary, digital code is harder to perform. Attempts are underway to overcome this technical hurdle. Graphene can be used in sports goods such as golf clubs, ski bindings, and so forth. Graphene sensors can be used for detecting any wavelength of light.

In order to fully utilize the advantages of graphene such as its high electrical conductivity and mechanical strength, chemical modification of graphene is needed. This will allow addition of the property specified. When biomolecules are attached to graphene, new sensors can be developed. Impurities in graphene would enable the material to rival the semiconductor behavior in electronic devices. It was found that resistance to functionalization of graphene was felt. Pressure was used in order to catalyze Diels–Alder reactions on the graphene surface in order to produce patterns of covalent modifications by scientists at the University of Miami and the University of California Los Angeles (UCLA). Cyclopentadiene, for example, is the Diels–Alder reactant. A 2×3 array of 80-nm-wide polymer tips was coated with inks containing compounds linked to cyclopentadiene. The tips were mounted on an AFM and were gently pushed onto the surface of graphene. Cyanine dye 3, for example, was attached to cyclopentadiene. Graphene was allowed to react with cyclopentadiene resulting in the formation of 20- to 40-μm patches of graphene decorated with 2×3 patterns of dye dots. The scientists performed the reactions at room temperature and pressure. They are working on functionalizing graphene with carbohydrates in order to make biological sensors.

Summary

Graphene has been identified as a distinct allotrope of carbon in addition to diamond, graphite, fullerenes, C_{60}, and CNTs. Graphene can be formed by unscrolling CNTs. Graphite comprises of graphene layers with an interlayer bonding strength of 5.9 kJ mol^{-1}. Graphene has a unique two-dimensional hexagonal lattice structure made up of sp^2-hybridized planar sheets of carbon atoms. It is superstrong, comprises one layer of honeycomb-structured atoms, and can be the next big discovery since James Watt's steam engine and John Bardeen's transistor and superconductor. It possesses interesting electronic, optical, mechanical, and thermal properties. A number of interesting applications are expected for single-layer graphene in the areas of computing, energy, and medicine.

The European Union is investing €1 billion as funding for 10 years in order to explore commercial applications of graphene. The Russian initiative is to spend $8.55 billion in order to create a nanotech industry by the year 2015. Korea has earmarked a $350 million initial amount in graphene development. There are 7,500 patents related to graphene. China, the United States, and South Korea are leading in the number of patents acquired on graphene. Top industrial participants in the graphene story are IBM, Xerox, and Samsung. €10 million is invested in Germany on a carbon innovation center. One hundred papers were presented at an APS meeting at Denver, CO, in 2007. The Nobel Prize for Physics was awarded in 2010 to A. K. Geim and Novoselov. Prof. R. Ruoff at the University of Texas, Austin, rubbed tiny pillars of graphite against a silicon wafer surface, causing them to spread out like a deck of cards. Discovery of graphene has sparked a global scientific gold rush.

TEM of graphene X is shown in Figure 1.1. Electrons are delocalized in the hexagonal sheet of atoms and move without any obstacle as shown in Figure 1.2. Chiral, armchair, and puckered morphologies of graphenes are possible. In order to puncture a graphene sheet of the thickness of a saran wrap, an elephant balanced on a pencil is needed. Graphene is a natural substance. It is 200 times stronger than steel and has a tensile strength of 1.5 million psi. International standards are yet to be established for graphene.

According to a recent Lux Research report, the projected market value of graphene by 2018 is $180 million. According to the BBC, by 2020, the market value of graphene will be $675 million. The Lux Research report did not include an economically scalable model of fabrication of graphene in its estimates. A number of scalable methods to make graphene are discussed in Chapter 5. The cost of production of graphene is expected to come down as the technologists move past the learning curve. It costs $60 per square inch of graphene on a copper substrate. Expectations are high for the costs to come down to $1 per square inch for industrial electronic applications and 10 cents per square inch for use in touch-screen displays.

Applications of graphenes are discussed in Chapter 3. They are of a wide range: high-capacity electrodes, antireflection coatings in solar cells, carbon composites for light-weight BMWs, panel displays in wireless telephones and laptops, thermal management, cancer treatments, featherweight HD televisions, inks, and NEMS. Graphene can be used in the study of sequences of RNA and DNA. von Neumann bottleneck can be obviated by designing novel MLGs. Bilayer graphene can be used to provide tunable bandgaps needed in supercapacitors, LEDs, and other applications.

CHAPTER 2

Characterization

Chapter Objectives

- SAXS, small-angle X-ray scattering
- TEM, transmission electron microscope
- Surface electron microscope
- SPM, scanning probe microscope, applications
- Microwave spectroscopy
- Raman spectroscopy
- STM, scanning tunneling microscope
- AFM, atomic force microscope
- Auger electron microscopy
- HeIM, helium ion microscope
- XRD, X-ray diffraction

Overview

Characterization of nanostructures such as single-layer graphene (SG) sheets needs suitable instruments. The resolution limits of optical microscopes cannot be greater than the order of magnitude of wavelength of light. As per the Raleigh criterion,[1] the resolution limit using optical devices is 200 nm—half the wavelength of light. X-ray devices are needed for studying nanostructures. Table 2.1 shows the resolution sizes of different instruments. A helium ion is the smallest ion and a helium ion microscope (HeIM) has a greater resolution power compared with a scanning probe microscope (SPM), which in turn has a greater resolution power compared with a transmission electron microscope (TEM), which in turn has a greater resolution power compared with a scanning electron microscope (SEM), which in turn has a greater resolution power compared with optical microscopes.

More about the equipment needed to characterize nanostructures is discussed in this chapter.

Raman Spectroscopy

The Raman microscope is designed using the principle of Raman effect. Sir C. V. Raman won the Nobel Prize for Physics in 1930 for his work on molecular scattering of light. He explained why the sky and seas were blue in color. He showed that molecules scatter light. The scattering depends on the characteristics of the molecule. Raman microscopes have been made to decipher the molecular type from the scattering information. Molecular scattering of light was primarily due to molecular disarray in the medium and due to the local fluctuations of optical density. The effect could be seen in vapors and gases and in crystalline and amorphous solids.

Raman effect occurs when a molecule is excited by light. The light interacts with the electron cloud of the bonds of that molecule. The incident photon excites one of the electrons into the virtual state. The molecule is excited from the ground state to a virtual state and relaxes into a vibrational excited state. This generates Stokes Raman scattering. The scattering can be used as a *fingerprint* from which a molecule can be identified. The changes in chemical bonding can be studied such as those when a substrate is added to an enzyme. Raman spectroscopy can be used to measure temperature and find the crystallographic orientation of a sample.

Raman spectra can be used to characterize single- and multiple-layer graphene samples. D* Raman scattering mode at ~2,715 cm^{-1} can be

Table 2.1 Resolution limits of different microscopes

10^{-11}	10^{-10}	10^{-9}	10^{-8}	10^{-6}	10^{-4}	10^{-2}	1
Helium ion microscope							
	Scanning probe microscope						
		Transmission electron microscope					
			Scanning electron microscope				
				Optical microscope			

used to identify the change in electronic structure between single- and multiple-layer graphene samples. As the sheet thickness increases, the narrow and symmetric peak becomes asymmetric on the high-energy side. Typical Raman spectra of graphene samples are shown in Figure 2.1.

An upshift of the G mode from ~1,580 cm^{-1} can be used to detect charge doping or compressive strain. Raman D mode at ~1,350 cm^{-1} can be used to detect with a sensitivity of 10^{11} cm^{-2}.[2] Scrolled graphene and folded graphene can be detected using Raman spectra. Podila et al.[3] examined the effect of incomplete folding in bilayer graphene by analyzing the shift in the G-band frequency. They showed that the different Raman shifts reported for the G-band frequency in folded SG can be the result of the relative strength of fold-induced electron/phonon renormalization. They found that the Raman spectrum of scrolled graphene differs significantly from that of flat graphene. They observed an enhanced D band in scrolled graphene due to momentum conservation of the intravalley

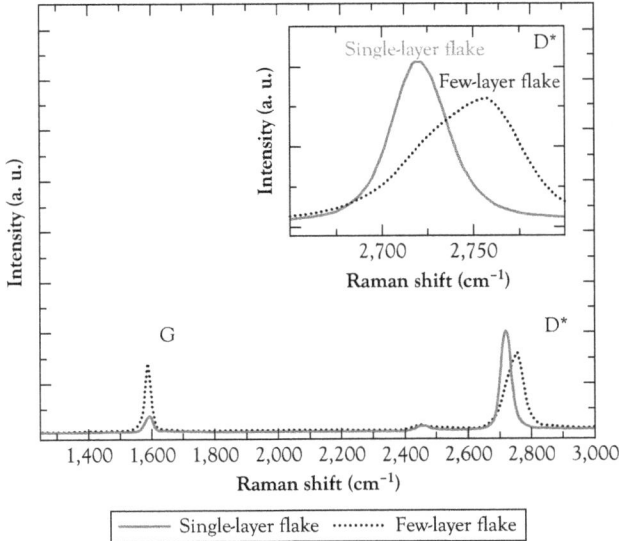

Figure 2.1 Prototypical Raman lines for single- and multiple-layer graphenes

Source: Stolyarova et al. (2007). High-resolution scanning tunneling microscopy imaging of mesoscopic graphene sheets on an insulating surface. *Proceedings of the National Academy of Sciences of the United States of America* 104 (29), 9209–9212. With permission from American Chemical Society.

electrons scattered via indium tin oxide (ITO) phonons (at the K-point) by curvature-induced defect scattering. In addition, they found that the curvature present in the scroll lifts the degeneracy between the longitudinal optical (LO) and transverse optical (TO) phonons near the K-point and leads to a splitting of the G band into three peaks located at 1,570, 1,580, and 1,592 cm^{-1}. Low-frequency radial breathing-like modes were used to confirm that the curvature in the scroll activates new modes that are absent in folded or flat graphene.

Intel has patented a microfluidic apparatus and methods for performing molecular reactions with nucleic acid molecules and Raman microscopic systems to detect these molecules.[4] The determination of the human genome sequence in its entirety has led to progress in identifying the genetic basis of diseases such as cancer, cystic fibrosis, sickle cell anemia, and muscular dystrophy. Current methods for measuring nucleic acid sequence information are tedious and expensive and oftentimes inefficient. Lack of harmonization in multimolecule polymer reactions such as that in the case of exonuclease sequencing of DNA results in imprecise results. An improved apparatus and method for performing biomolecular reactions were developed by Intel. It allows for the capture of a single nucleic acid molecule in a microfluidic channel upstream from an optical detector. *Sequential detection* of one or more nucleotides is made using a Raman microscope. Optical tweezers are used to isolate a single nucleic acid molecule. Lasers that are used by optical tweezers may interfere with the detection using a Raman microscope. Integration of optical tweezers and Raman microscope may limit the field of view. The Intel apparatus does not allow for interference of Raman microscope detection capabilities. Solid supports such as particles or beads can be used to allow for immobilization of a single nucleic acid molecule on them. An optical tweezer is usually a gradient force optical trap such as a single-beam gradient force optical trap that captures the single particle downstream from the laser beam. Single nucleotides can be cleaved from the bead using an exonuclease. Single nucleotide molecules are detected using *surface-enhanced Raman spectroscopy* (SERS). The inclusion of a restriction barrier in a microfluidic channel and the immobilization of an optically transported bead allows for removal of the optical tweezers from the optical path of the detection device. Thus, the interference from the additional light source of the optical tweezers close to the collection volume of the detector is obviated.

The SERS detection unit includes a detection light source, typically a laser light source used for irradiation of the molecule, and a detection unit for capturing Raman emission from the emitting molecule. Thus, the nucleic acid molecule attached to the surface of the particle is analyzed. The system can be used to obtain the *sequence distribution* of the nucleic acid molecule. The movement of the nucleic acid molecule is restrained. The molecule is then contacted with an agent that removes nucleotides such as an exonuclease. A terminal nucleotide is released and then detected using SERS. Raman emission from the first and second released nucleotides is detected. Polymerase chain reactions (PCRs) can be analyzed in a similar fashion using hybridization reactions, fluorescent probes, and Raman-labeled molecular probes.

The apparatus and system consist of: (1) light source; (2) detector to detect SERS emission of a molecule excited by the light source; and (3) first channel that provides a restriction barrier. The length of the nucleic acid molecule can vary from 10 base pairs to 5 million base pairs.

Helium Ion Microscopy

The helium ion microscope (HeIM) was developed to expand the capability of electron microscopes. Imaging is obtained using helium ions in place of electrons. Shorter wavelengths are associated with helium ions compared with electrons. More tightly focused beams are possible. Better image resolution is achieved in this manner. The physics of secondary electron generation and ion backscattering is used to collect image information that is not possible with a SEM.

Orion is the first commercially available HeIM and was installed in the National Institute of Standards & Technology from Carl Zeiss, Peabody, MA. It is based on a commercially manufactured helium field ion gun. The world record for resolution is broken by Carl Zeiss SMT Inc, Peabody, MA. TEM-like resolution on bulk samples with SEM ease can be achieved using an ORION PLUS microscope. A surface resolution of 240 picometers has been achieved reproducibly using an ORION HeIM. Atomic-level ion source (ALIS) technology is used to provide a stable helium ion beam with high brightness and hence a subnanometer probe size on the sample. The source is based on field ion emission from the apex of a sharp needle held at a high positive voltage in the presence of a gas. Field ion microscopes have been built using this phenomenon and used in advanced materials analysis. The electrostatic optics can be used to produce, in theory, a focused probe with a diameter of 250 pm.

The resolution is close to the diameter of a single atom. It is three times better than the SEMs with the same surface sensitivity. A new benchmark was achieved for surface imaging in the subnanometer range. Semiconductor manufacturers use the HeIM in order to observe small features that currently cannot be resolved. Advances in miniaturization of feature sizes of semiconductor devices require high-resolution microscopy as a must. Some layers of integrated circuits (ICs) have reached a thickness of only a few atoms. Images with Rutherford backscattered ions (RBIs) can be created using an HeIM. RBIs are high-energy helium ions that rebound off a sample. Information on chemical composition can be obtained using HeIM that cannot be obtained using SEM. The chemical composition of a defect in a semiconductor chip can be detected using HeIM.

The source of the microscope is small and the helium ions emanate from a region as small as a single atom. On account of the lower wavelength, the helium ions do not suffer appreciably from adverse diffraction effects. Adverse diffraction effects are a law of physics that predict a fundamental limit on the imaging resolution of electrons. Signals are triggered directly from the surface of the sample by the helium ion beam. They stay collimated upon entering the sample. As a result, sharp and surface-sensitive images at the quoted resolution can be obtained. A majority of secondary electrons that are used for imaging stem from deeper, less confined regions within the sample causing blurrier images with less resolution in a SEM compared with images from an HeIM. The cost of an HeIM in 2008 was $2 million.

HeIM is related to field-ion or field-emission microscopy. These were used for observing individual atoms on a cryogenically cooled tungsten tip in an ultrahigh vacuum system with small amounts of helium gas present in it. A gallium ion microscope can be used to analyze a wider range of samples. The problem of sputtering of the sample prior to imaging remains with HeIM. A distinct pyramid-shaped ion source allows helium ions to form a more tightly focused beam leading to images with better resolution. Helium ions are lighter than gallium ions and the samples are not allowed to deteriorate readily.[5] Helium as an ion source is the primary component of an HeIM. Images can be collected in two modes: the RBI and the secondary electron mode. A combination of heat and electric fields is used to allow for emission of electrons. The signals are synthesized into an image at the detector. Helium ions hit the sample. Upon a hit, secondary electrons and RBIs are released. Helium ions possess greater mass and much shorter wavelength compared with electron beams. Helium ions interact more strongly with the materials compared with electrons and 100 times more secondary electrons are produced. More information goes to the detector, resulting in images with greater details. Researchers in nanoscale materials are eager to use HeIM to obtain information on chemical composition from RBI images. Useful information about sample surface can be obtained using secondary electron images.

Automated microelectronic chip manufacturing industry can use HeIM for monitoring in place of SEM. One weakness of an HeIM is that more damage to the specimen is caused in it compared with that in electron microscopes. Properties of cellulosic nanocrystals and carbon nanotubes can be examined using HeIM. Chemistry of nanowires and other nanoscale materials was examined using HeIM at Harvard University, Boston, MA, using both the RBI and secondary electron modes.

Inspection of nanofabricated structures and nanomanufacturing can be effected using HeIM. Monolayers of graphene and images of dry biological samples can be studied using HeIM.

As the particle microscope is charged, interactions between the beam and sample are needed. Distinct contrast levels can be seen in HeIM images along grain boundaries. In SEM images, contrast between grain boundaries is minimal.[6] A greater range of gray levels is seen in HeIM images compared with that in SEM images. This is because the energy loss for an ion beam at this energy is dominated by nuclear–nuclear scattering rather than electron scattering. Surface chemical states can be investigated. Catalysts and graphenes can be analyzed using HeIM. Nanopatterning of graphene sheets with metals involves modification of surface states. Failures in electronics can be studied using HeIM by the resolution of whisker formation and growth.

Scientists at Harvard University have reported the etching of graphene devices using helium ion beams. They can also perform in situ electrical measurements during lithographic processing. Electrical isolation of different regions in the device and nanostructuring can be achieved using the etching process. A channel in a suspended graphene device was etched[7] with etch gaps down to 10 nm. When the substrate used was SiO_2, the etching was achieved at lower doses of He ions.

Small-Angle X-Ray Scattering

The nanoscale structure of advanced material systems can be studied using small-angle X-ray scattering (SAXS) devices in terms of relevant parameters. For example, the combined effects of graphene nanosheets and shear flow on the crystallization behavior of isotactic polypropylene (iPP) were investigated using SAXS techniques.[8] The X-ray source can be synchrotron light, which provides high X-ray flux. SAXS can be used in order to obtain information about the shape, size, internal structure, crystallinity, and porosity of nanostructured samples.

Structural information in the scale of *5 to 25 nm* can be characterized using SAXS. The test is nondestructive and crystalline samples are needed. A monochromatic source of X-rays is used to excite the sample and the scattered X-rays are detected by a two-dimensional flat X-ray detector. The structure is deduced from the *scattering pattern*. The weakly scattered beam needs to be separated from the main beam with large strength. The difficulty increases with the decrease in the desired angle. Divergent beams are produced from X-ray sources compounding the problem. Beam focus is used to overcome the difficulties encountered. Collimation is relied upon as beam focus is not readily accomplished. In point-collimation instruments, the X-ray beam is shaped by the use of pinholes to a small spot with a circular or elliptical shape in the plane of detection. Scattered intensity is expressed as a function of scattering vector in the observed data. *Porod's law* states that the contribution to the scattering that comes from the interface between two phases and the intensity should drop with the fourth power of q if this interface is smooth: $I = kSq^{-4}$. The surface area S of the particles can be determined by

SAXS. For a *fractally* rough surface area with a dimensionality d between 2 and 3, Porod's law becomes: $I = kSq^{-(6-d)}$. The particle size distribution can also be obtained from the SAXS data. The *Guinier approximation* can be applied to the beginning of the scattering curve at small q values. The intensity in this regime would depend on the radius of gyration of the particle. The distance distribution function is calculated by the use of Fourier transform. The distance distribution $p(r)$ is related to the frequency of certain distances r within the particle.

X-ray scattering has been used to study material structures that possess *long-range order*. The technique is based on elastic scattering of X-rays by the structures under scrutiny. For SAXS to be an effective technique, deviations from the average electron density have to be present in the sample. When these are concentrated in systems, the intensity in SAXS measurements is proportional to the Fourier transform of their form factor. The Guinier approximation can be used to show that the intensity variation with the scattering wave vector close to the origin is the same regardless of the shape and depends only on the size of the particle. An additional contribution from particle to particle correlation can be included in agglomerated systems and $I = F(g)S(q)$.

The structure factor is the measure of short-range order. A maximum is displayed at an angle reciprocal to the average of the first neighbor distance. SAXS use is also limited when thin-layered systems on thicker substrates are used. The deep penetration depth and greatly reduced signal-to-noise ratio are the limiting attributes of the instrument. When grazing incidence is used, the technique is called GISAXS. When grazing angle is varied, the penetration depth can be limited. This can provide information up to several 100 nm into the substrate, which was not accessible by other microscopy techniques.

SAXS was shown to be versatile[9] in characterizing nanostructures. It is a nondestructive technique. It can be used for one-dimensional structures such as thin films or thin multilayered systems or three-dimensional oriented nanoparticles of different chemical compositions either buried in a thin film or grown on a substrate surface. It is more advantageous over near-field microscopy techniques. It gets selected for material characterization in the 1 to 100 nm range.

SAXS is rapidly becoming a well-established analytical method for nanostructure analysis. When X-rays are allowed to penetrate the specimens, they get scattered at the interfaces of nanostructures. A scattering pattern is produced that is specific to the nanostructure. The method is nondestructive, less expensive, and requires less sample preparation. It allows for investigation of interactions between molecules in real time. These interactions may lead to self-assembly and large-scale structure changes on which material properties depend upon. Its applications have been found in a variety of fields such as emulsions, liquid crystals, and macromolecules to porous molecules and metal alloys. It finds use in research, technology development, and statistical quality control.

Nanoscale particles scatter toward small angles. The SAXS pattern provides information on the overall size and shape of these particles. The orientation and nanostructure of the sample can be obtained. Atoms and interatomic distances scatter toward large angles. The obtained wide-angle X-ray scattering (*WAXS*) pattern provides information on the phase state, crystal symmetry, and the molecular structure.

True SWAXS is a feature of SAXsess that allows for simultaneous measurement continuously from small and wide angles upto 40° with a consistent high resolution and without changing the instrumental setup. The complete information on the nanostructure of the sample and its phase state such as crystalline or amorphous can be obtained in a single run. A high-intensity signal is produced at the detector. Nanostructures in the range of *0.2 to 150 nm* can be investigated. The ease of handling of the facility aids the measurement procedure and provides for ready alignment. Line collimation is offered for rapid data acquisition of isotropic samples and point collimation is provided for studies of oriented samples in one device. Detection systems provide remarkable resolution. The device consists of 10 different components such as X-ray source, advanced focusing multilayer optics, enhanced block collimation system, sample stage, semitransparent beam stop, TrueSWAXS—single-run feature, high-performance detector system, SAXSsquant software system, temperature control, and versatile sample holders.

SAXSpace is shown in Figure 2.2. This is based on small- and wide-angle X-ray scattering (SWAXS). X-rays are scattered toward small angles for studying nanosized particles. The scattering pattern provides

Figure 2.2 SAXSpace from Anton paar for nanostructure analysis

information on the overall size and shape of the nanostructures. It provides for easy handling and push-button alignment and versatile and precise sample stages. SAXSpace can be used for observing nanostructures in a host of different materials such as proteins, foods, pharmaceuticals, polymers, nanoparticles, and nanostructured surfaces. The scatterless beam collimation concept is used in order to obtain high-resolution and precise SWAXS data.

Transmission Electron Microscope

Transmission electron microscopes (TEMs) can resolve structures in the nanoscale region. Sample preparation for TEM studies is complex and requires other instruments. The sample specimens need to have a thickness down to *few hundred nanometers* depending on the operating voltage of the instrument. They must possess parallel surfaces. A thin section of 0.5 to 3.0 mm is cut from the bulk material using electric discharge machining or other such similar techniques and a rotating wire saw. The specimen is milled down to 50 μm thickness. Electropolishing and ion beam thinning are used to finish the sample to its final dimensions.

A high voltage of 100 to 300 kV is applied to a tungsten filament and an *electron beam* is produced. This is then accelerated down to the specimen. Electromagnetic coils are used to have the electron beam condensed. The electrons are allowed to pass through the specimen. As the electrons pass through the specimen, some are absorbed and get scattered and change their direction. Sample thickness is a critical parameter. With thick samples due to excessive absorption and diffraction, the transmission of electrons will not be permitted. Electron scattering is caused by differences in atomic crystal arrangement. The electron beam upon transmittal is focused with a magnetic lens and then amplified and projected on a fluorescent screen. An image is formed that is either a bright-field image or a dark-field image depending on whether the direct beam or the scattered beam is selected. Irregular atomic arrangement such as dislocations in material imperfections will appear as dark lines on the electron microscope screen.

A high-resolution transmission electron microscope (HRTEM) can achieve resolution sizes down to 1 A°. Atomic-level phenomena can be viewed using HRTEM techniques. VG Microscopes have developed scanning transmission electron microscopy (STEM) instruments. The objective lens in theses instruments is a strong *electromagnetic lens*. This allows for demagnification of the electron source and formation of a fine probe of the specimen. The excitation source is of the cold-field emission type for minimum size and maximum brightness. Noise is reduced by operation in a vacuum of upto 10^{-10} Torr. A 100-kV source offers the energy for acceleration of the electrons. One to three condenser lenses precede the objective lens. The beam configuration and apertures may be selected depending on the application. The fineprobe is allowed to be scanned over the specimen. When the fine probe is passed through the specimen, a convergent beam diffraction pattern is formed on the distant plane of observation. This may be observed and recorded by using a suitable phosphor screen and charged coupled device (CCD) cameras. The STEM signal is formed from a part of this pattern. This is then displayed on a cathode ray tube with a scan synchronized with that of the probe at the specimen level to form the magnified image. A bright-field image is formed by detection of a portion of the central beam of the convergent beam electron diffraction (CBED) pattern. Dark-field images may be obtained by detecting individual diffraction spots. Electrons scattered outside the central spot may be collected using an annular detector.

The specimen is placed within the magnetic field of the objective lens for high-resolution STEM imaging. A focusing effect is produced by the magnetic field of the objective. A lens system is formed by using two or more post-specimen lenses. One advantage of STEM instruments compared with TEM instruments is the design flexibility.

TEM has been used in life sciences and biomedical investigation because of its ability to view the finest cell cultures. It is used as a diagnostic tool in pathology laboratories of hospitals world over. High-voltage/high-resolution TEMs manufactured by JEOL, Tokyo, Japan, utilizing 200 keV to 1 MeV, offer resolution sizes down to the *size of the atom*. This allows for imaging of atoms and design of materials with tailor-made properties. TEM can be used as an elemental analysis tool with the addition of energy dispersive X-ray analysis (EDXA) or electron energy loss spectroscopy (EELS). Elements in areas less than 500 nm in diameter can be identified using this tool.

In Figure 2.3 an HRTEM image of SG is depicted.[10] This image is used to arrive at the number of layers of graphene sheets and whether multiple folds exist. The graphene shown in Figure 2.3 was produced by

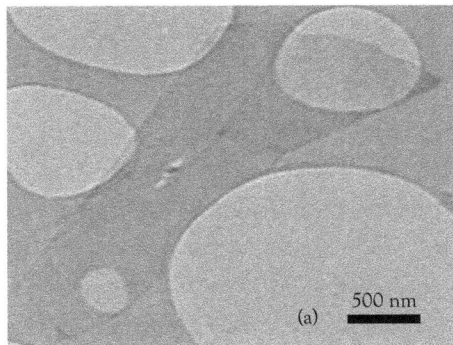

Figure 2.3 An HRTEM image of SG

dispersion and exfoliation of pure graphite in *N*-methyl-2-pyrrolidone (NMP). The monolayer yield was found to be 1 wt% and could be improved to 12 wt%. The energy required to exfoliate graphite into a single layer was overcome by solvent–graphene interactions. The solvent has similar surface energy compared with graphene. This holds the potential for large-scale production of graphene.

A low-resolution cryo-TEM image of surfactant peptides at pH 7 at a peptide concentration of 5 mg mL^{-1} was reported in a recent patent from MIT.[11] They discussed dipolar *oligopeptides* that self-assemble to form regular structures. Gold is localized upon the nanostructures that are formed by self-assembly. Different nanoarchitectures are formed by self-assembly. Short oligonucleotides with di- and tri-block peptide copolymers with properties that mimic those found in surfactant molecules were synthesized. Well-defined structures of about 50 nm were prepared by self-assembly of short amphiphilic peptides. Such *self-assembled structures* can be modified by external parameters such as pH. Appropriate designing of the peptide allows fine-tuning of the self-assembly properties and offers flexibility for various applications. One class of peptides has been designed and investigated for its ability to self-assemble spontaneously to form stable nanotubes. It consists of hydrophobic and hydrophilic groups. The lipophilic tail is made up of alanine, valine, isoleucine, or leucine. The hydrophilic head is made up of charged amino acids such as lysine, arginine, histidine, aspartic acid, and glutamic acid. When dispersed in water, the amphiphilic peptides tend to self-assemble, leading to the formation of a polar interface that separates the hydrocarbon and water regions.

Cryo-TEM is performed at temperatures of −160 to 50°C. An electron beam is transmitted into the specimen and focused on it. The image contrasts are formed by scattering of electrons out of the beam and various magnetic lenses are allowed to perform *in lieu* of ordinary lenses in an optical microscope. The diameter of the oligopeptide sequence is about 10 to 15 nm. In order to obtain an aqueous solution of the oligopeptides, it was found necessary to deprotonate the carboxylic groups by changing the pH with a solution of 0.1 N NaOH. Solubilization was found to begin at a pH of 5 to 6 depending on the amino acid sequence. It was found from cryo-TEM investigation that charged oligopeptides exist as a dense network of entangled nanotubes with diameters ranging from 25 to

50 nm. Cylindrical assemblers provide a three-dimensional transient network. Only two-dimensional projection of the peptide nanotubes were found to be imaged. Cylindrical morphologies were found for the V_6D oligopeptide structure (V is valine and D is aspartic acid). High axial ratios were found in the self-assemblies. The length extends to several microns. Many threefold junctions or branching connect the nanotubes forming the final network. This kind of branching can be shown to be energetically unstable.

Branched supramolecular structures have been discussed in the literature. Branching has been found in aqueous surfactant solutions. *Reverse structures* such as lecithin organogels are also found to form threefold junctions. Patches are produced at branch points having a mean curvature opposite to that of the portion far from the junction. Such branching points have been attributed as causative in the manifestation of viscoelastic properties of polymer-like systems. Branches versus *entanglements* have been studied. Thus, a tubular nanostructure can be constructed from self-assembly of oligopeptides. Fine-tuning of monomer properties will give rise to a wide range of nanostructures. These can result in the development of novel biomaterials.

Scanning Electron Microscope

Using electron channeling contrast, M. Knoll in 1935 obtained the world's first SEM image of silicon steel. It was first marketed in 1965 by an outfit named Cambridge Instrument Company in the name of Stereoscan. Magnification in a SEM is over five orders of magnitude ranging from 25X to 250,000X. In SEMs, the image magnification is no longer a function of the objective lens. The electron beam is focused on a spot. As the *electron gun* is made to generate a beam with sufficiently smaller diameter, the need for condenser and objective lenses ceases. Similar to SPM, the magnification results from the ratio of the dimensions of the raster on the specimen and the raster on the display device. The current supplied to the scanning coils determined the magnification.

The *spatial resolution* of the SEM depends on the wavelength of electrons and the electro-optical system that produces the scanning beam. The extent to which the material interacts with the electron beam is also an important consideration. Resolution of the atomic scale is not possible as can be seen in the case of TEM. However, the advantage of using SEM is its ability to scan a larger area of the specimen and to image bulk materials. A variety of analytical modes are available for determining the composition and properties of the specimen. The resolution of SEM can range from less than *1 nm to 20 nm*. Hitachi S5500 offers the highest SEM resolution world over and is *0.4 nm* at 30 kV and 1.6 nm at 1 kV. User friendliness is higher for SEM images compared with TEM images.

A typical SEM consists of an electron gun that is capable of emitting electrons thermionically. The electron gun is fitted with a tungsten filament cathode. The electron beam thus generated possesses energy values of 40 keV to a few hundred electron volts. It is focused by condenser lenses to a spot of about 0.4 to 5 nm diameter. The scan is made in a raster fashion over the area of the sample surface. The electron beam is allowed to interact with the sample within an interaction volume of 100 nm to 5 µm. The loss of energy from the electrons is by repeated random scattering and absorption within the interaction volume that is tear-drop shaped. Elastic scattering, inelastic scattering, and backscattering of electrons take place.

The SEM images the sample surface by *raster scanning* using electrons. The surface *topography, composition,* and other properties can be obtained from the scan. Samples must be of a suitable size that can fit into the specimen chamber. The specimen must be electrically conductive. Little sample preparation is needed. Iridescent biological samples are sputter coated with gold for preparation of SEM imaging. The backscattering and quantitative X-ray analysis require the specimen to be ground and polished to an ultrasmooth surface.

An SEM image of one-dimensional silver-nanowire-doped graphene is shown in Figure 2.4. This is used in conductive and flexible paper applications of graphene. Chen et al.[12] proposed a novel architecture of a graphene paper that can comprise metallic nanowires in one dimension in defect-free graphene sheets. A simple filtration method was used in order to fabricate a material with increased conductivity and improved flexibility with silver nanowires and graphene sheets made by chemical vapor deposition (CVD). Graphene papers made by CVD have been found to have a higher electrical conductivity of 1,097 siemens cm^{-1}. Further

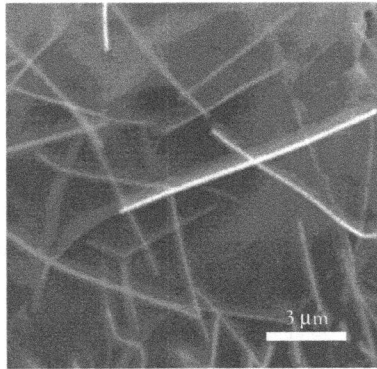

Figure 2.4 An SEM image of nanowires (distance captured: 300 μm)

improvement of electrical conductivity to 3,189 siemens cm^{-1} can be achieved by the addition of silver nanowires. Graphene papers offer flexibility to the extent of 500 times mechanical bending. Such composite papers can be used in high-performance, flexible energy conversion and energy storage devices.

Nanowires are found to be oriented longitudinally in a single direction along the direction flow during the deposition process. There is increased interest in semiconductor nanowires on account of their interesting and novel electrical, chemical, and optical properties. They are used in nanolasers, photovoltaics, and sensor applications such as nano-Chem-FETs (field-effect transistors). Positioning of these materials has been a technical challenge.

Nanosys patented a method for preparing *oriented nanostructures*.[13] Nanowires are deposited on a surface substantially in a desired orientation. A fluid containing nanowires is made to flow over a surface. The nanowires are then immobilized on the surface with the longitudinal dimension of the nanowires oriented in the first direction.

Scanning Probe Microscope

The development of an SPM is part of a revolution in characterization and analysis of materials at the nanoscale. These instruments are different from the optical and electron microscopes. Neither light nor electrons is used to form the image. A topographical map is generated on the atomic scale. The surface features and characteristics of the specimen being studied are obtained. Magnification higher than *1 billion* is possible. Nanoscale features can be examined using SPM techniques. Better resolution is possible with these methods. Three-dimensional images that are amplified are obtained that contain topographical features of interest. These instruments may be operated in vacuum, air, liquid, or any other desired environment.

A *tiny probe* with a sharp tip is used and is brought in close proximity within 1 nm of the specimen surface. Across the plane of the surface, the probe tip is then raster scanned. Deflections perpendicular to the plane are experienced in response to electronic interactions between the probe and the surface. The probe movements in and out of the surface are controlled by piezoelectric ceramic components with nanometer resolutions. The motion of the probe tip is also recorded electronically and displayed in the computer monitor using suitable data acquisition techniques. A three-dimensional surface image is then generated. Vacancy defects can be captured using SPM. An atom missing from an otherwise regular lattice can be imaged using the SPM technique. Biomolecules and silicon microprocessors have been examined successfully using SPM. It aids in engineering of nanomaterials.

SPM can be used to prepare nanostructures. Dip pen nanolithography (DPN) uses the SPM tiny probe tip. Dip pen lithograph techniques have been known from ancient times. They are about 4,000 years old. Ink on a sharp object is transported to a paper substrate by capillary forces. DPN can be used to *write patterns* consisting of only a collection of molecules. An elastomer stamp may be used to deposit patterns of thiol-functionalized molecules directly onto gold substrates. In this technique, molecules are delivered to a substrate of interest in a positive printing mode. The solid substrate is used as the "paper" and the SPM tiny probe tip is used as the "pen." The probe tip is coated with a patterning compound, the "ink." The desired pattern is produced by application of the patterning compound to the substrate. Capillary transport is the mechanism of delivery of the molecules of the patterning compound to the substrate. A variety of nanoscale devices can be produced using this technique. Software can be developed for computer-controlled performance of DPN. Many SPM tips are available in the open market. Park Scientific, Digital Instruments, Molecular Imaging, Nanonics Ltd, and Topometrix are some examples of vendors of SPM tiny probe tips.

As an alternative, SPM tiny probe tips may be tailor made to suit the needs of the given application. They can be prepared using e-beam lithography. A solid tip with a hole bored in it can be made using *e-beam lithography*. SPM tips can be used as atomic force microscope (AFM) tips. The patterning compound gets attached to the substrate by *physisorption*. It can be removed from the surface using a suitable solvent. The physisorption may be enhanced by coating the tip with the adhesion layer and by judicious choice of solvent. The adhesion layer is less than 10 nm in thickness. This layer should not change the tip's surface. The strength the layer can withstand AFM operation is upto 10 nN. Examples of materials that make good adhesion layers are titanium and chromium. Northwestern University (http://www.anton-paar.com/-#http://www.anton-paar.com/-#)[14] has patented a method to use an SPM tiny probe tip in DPN to prepare nanostructures.

The SPM tiny probe tip was coated with 1-octadecanethiol. A multi-staged etching procedure was described to deposit alkylthiols onto a gold–titanium–silicon substrate. Alkylthiols form well-ordered monolayers on gold thin films that protect the underneath gold from dissolution during certain wet chemical etching procedures. This is true for DPN-generated resists. The gold, titanium, and silicon dioxide that were not protected by the monolayer could be removed by chemical etchants in a staged procedure. This procedure yields first-stage three-dimensional features and multilayered gold-topped features on the silicon substrate. Second-stage features were affected by using the leftover gold as an etching resist to allow for selective etching of the exposed Si substrate. The remaining gold was then removed to yield final-stage, all-silicon features. DPN can thus be combined with wet chemical etching to obtain three-dimensional features on Si (100) wafers with at least one dimension in the sub-100-nm scales. Nanoscale features can thus be obtained on Si wafers. It starts with the coating of 5-nm titanium onto polished single crystalline Si (100) wafers. This is followed by coating of 10 nm of gold by thermal evaporation.

In addition to surface topography data, many other data sets can be obtained using scanning probe microscopy (SPM) such as: (1) using scanning tunneling microscopy (STM), the conductance and current/ distance measures can be obtained; (2) AFM is used for measurement of lateral force and adhesion; (3) near-field optical microscopy (NFOM) is used for laser transmission at various wavelengths and polarizations; and (4) magnetic force microscopy (MFM) is used for measurement of temperature and parameters. At the University of Tokyo, a direct coupling system between the nanometer and human scale worlds was used. A stereo SEM is augmented with two-handed force feedback control. The sample is positioned within the SEM in the left-hand position and the probe is sampled with a tool by the right hand. A magic wrist was connected to the STM at IBM. The STM translated across the surface the user movements in x and y. The device was kept above the surface. The surface of the gold could be felt by using this technique. System vibrations were noted as a technical problem.

The formation of graphene on SiC-OI (on-insulator) substrates was investigated using STM. Graphene-on-insulator structures obtained on

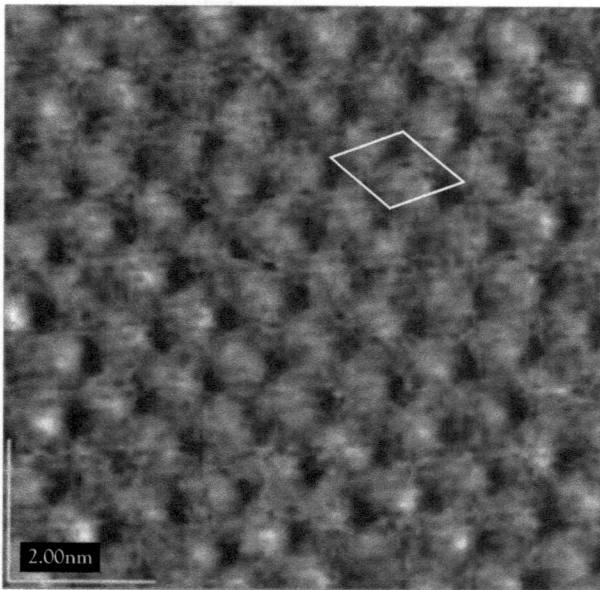

Figure 2.5 Investigation of formation of graphene using STM: STM image after annealing 3C-SiC (111) on SiO$_2$/Si (111) for 5 min at 950°C

these substrates can be used in making graphene-based nanodevices. Annealing (Figure 2.5) of an SiC-OI substrate with an SiC thickness of 1,500 nm was used in order to produce a graphene layer on the SiC surface. When the thickness of SiC was reduced to 5 nm, the graphene layer was not formed. The preparation of graphene involves the surface decomposition of SiC. Heating the Si-terminated hexagonal face corresponding to an SiC (0001) or SiC (111) surface at high temperatures results in sublimation of Si by thermal decomposition of the SiC surface, and graphene layers are formed parallel to the surface of SiC.[15]

Microwave Spectroscopy

Phenomena such as Coulomb blockade and single-electron tunneling were detected when transport was allowed in small electronic islands in semiconductor nanostructures. The dimension of these devices is 500 nm.[16] They contain 10 to 100 electrons and are referred to as *quantum dots*. Transport through these devices was found to be an oscillating pattern of the conductance through the dot as a function of gate voltage. Quantum dots can be used as photon detectors. Measurements of photocurrent induced through single and double quantum dots have been reported. The millimeter wave radiation in the range of 30 to 200 GHz was coupled to different antenna types. Photon-assisted tunneling (PAT) through the quantum dots can be employed for spectroscopy. *Rabi oscillations* were found to occur when two dots were strongly coupled. When the electromagnetic field interacts with the oscillating valence electron, the oscillations can be probed directly by time-dependent measurements.

The influence of high-frequency microwave radiation on single-electron tunneling through a single quantum dot was investigated. Tunneling of electrons is allowed. Coupling between radiations to the quantum dot is affected by an on-chip integrated broadband antenna. An additional resonance was attributed to PAT. The detection principle is based on absorption of photons by electrons in the leads leading to the PAT. The energy acquired by electrons allows for tunneling above the Fermi energy state. The mechanism of tunneling between two superconductors separated by an insulator is used for high-frequency detection. Application of high-frequency radiation with frequencies greater than

100 GHz results in the modification of the Coulomb blockade oscillations. Using electron spin resonance (ESR) in a single quantum dot at filling factor $v \leq -2$, Lamb shift was observed in hydrogen. The effect of electron spin on electron transport through a quantum dot is considered. The quantum dot structure is then imaged using an SEM.

PAT can be used as a tool for millimeter wave spectroscopy. Measurements using microwave radiation lent credence to the finding that the effect of Coulomb blockade can be overcome with the continuous wave radiation. Time-dependent measurements can be used to obtain details about electronic modes in quantum dots. In quantum dot systems, studied so far, only continuous wave (CW) frequency sources have been applied for spectroscopy in the range from some MHz to 200 GHz. Recent works reported in the literature demonstrated the ability of integrated broadband antennas to couple CW millimeter and sub-millimeter radiation to nanostructures such as quantum dots and double quantum dots. The expected PAT resonances in transconductance are shown. The energy acquired by the electrons through absorption of the photons allows for tunneling through states of higher energy above the Fermi energy.

Thus, millimeter wave spectroscopy can be used on quantum point contacts and single and coupled quantum dots. The influence of potential asymmetries in such devices has been discussed in the literature. Photon-assisted transport through single and double quantum dots in the high-frequency regime has been demonstrated. This method can be applied for spectroscopy of quantum dots far from equilibrium. This was demonstrated using ESR for a single dot at high magnetic fields. A millimeter wave interferometer was used to perform coherent spectroscopy on coupled quantum dots. The whole millimeter wave regime is covered and allows for both magnitude- and phase-sensitive detection. The high-frequency conductance through the double quantum dot was detected using the measurements shown. A broadband response was found. Strong vibrations in the relative phase signals were observed when the frequency of radiation was larger than the Rabi frequency of the "artificial molecule." The coherent mode in the double quantum dot is destroyed.

Liu et al.[17] developed a microwave "Corbino" spectrometer in order to measure the broadband response in the frequency range of 100 MHz to 16 GHz of CVD-grown graphene at temperatures down to 330 mK.

Broadband spectral information in the microwave regime was obtained. Data for both sheet impedance and complex conductance were presented. The measurement of the intrinsic impedance of a single atomic layer on an insulating substrate offers a number of technical challenges. They found two different scattering rates in the microwave and time domain terahertz spectroscopy measurements. They attribute this difference to the limitations of the Drude model for electron transport in graphene at low frequencies. Extracting complex impedance of graphene was difficult. Graphene conductance can be isolated from the substrate impedance by going to the high-bias/large-conductivity regime.

Auger Electron Microscopy

A scan using an *Auger Electron Microscope* identifies the elemental composition of the analyzed surface. Multiplex scan quantitates the atomic concentration of the elements identified in the survey scans. Detection limits are one-tenth of 1% of the atomic composition of the elements. AEM mapping is used for measurement of lateral distribution of elements on the surface. The spatial resolution is about 300 nm. Depth profile measurements of distribution of elements as a function of depth into the sample can be obtained. The depth resolution depends on the sample and sputter parameters. Less than 10-nm resolution is possible at a typical sputter rate of 3 nm min^{-1}. The sample must be conductive and prepared appropriately. The sample must also be compatible with a high-vacuum environment. The typical analysis time is 30 min per sample. Reproducibility is within 10% of the relative error. Capabilities include measurement of elemental composition of small conductive areas, particles, inclusions, and semiconductor devices.

Auger spectroscopy may be used for chemical specification and chemical bonding characteristics of the material and any adsorbate present before or after exposure to blood in the case of biomaterials used as artificial organs. Biomaterials are synthetic materials used to replace a part of human anatomy. It has to function in contact with living tissues. Cardiovascular diseases cause a major portion of deaths in the nation. Continuous improvements in the development of biomaterials capable of substituting for parts of the cardiovascular system are important.

Examples where implants are used are heart valves, prostheses, stents, and vascular grafts. Some of the requirements for materials used to prepare these implants are biocompatibility, thrombresistivity, nontoxicity, and durability. A key technical hurdle in an interfacing biomaterial with blood revolves around the characteristics of the implant surface. Thrombus and embolism formation at the blood implant interface is a primary concern. Ceramic implants provide chemical inertness, hardness, and wear resistance. They lack the ability to deform plastically.

A multilayered protective coating was formed of ceramic materials. The thickness of the coating layer was in the range of 1 to 100 nm.[18] The coating comprised of an inner component with one or two layers of zirconia, titania, or alumina. It also contained an outer component formed using a water-swellable ceramic material capable of forming a hydrate or hydroxide compound upon contact with an oxygen-containing environment. The outer component is made of aluminium, zirconium, or halfnium. The overall thickness of the coating is several microns.

Atomic Force Microscopy

The STM evolved into the AFM. AFM is a special type of SPM, discussed in the section "Scanning Probe Microscope." The resolution achieved using an AFM is a fraction of a nanometer.

The AFM was invented in 1986 and is used in order to characterize nanostructures. The specimen surface is scanned using a microscale cantilever with a probe at its end with a sharp tip. The tip radius of curvature is of the order of a few nanometers. The deflection of the cantilever is caused by the forces between the tip and the sample surface. The deflection is found to obey Hooke's law of elasticity. Different types of forces can be measured using the AFM, such as van der Waals forces, hydrogen bonding forces, surface tension forces, magnetic forces, electrostatic forces, and so forth. The laser source excites the specimen surface. The cantilever reflects the laser light and is captured by the array of photodiodes. Cantilevers are made up of piezoresistive elements such as strain gauge, and a wheatstone circuit is used to measure the deflection. The tip-to-sample distance is controlled using a feedback control loop. AFM can be operated in the imaging or tapping mode. Individual atoms can be imaged using AFM. Atoms of silicon, tin, and lead on an alloy surface can be distinguished using AFM.

AFM has been used to form complex patterns in thin layers of MoO_3 grown on the surface of MoS_2.[19] The pattern lines formed by using AFM were found to be less than 10 nm. The resulting structure is imaged using the AFM as well. The AFM has been used as a tool to move atoms or clusters of atoms directly into a desired configuration. Orientation ordering of organics can be effected by direct contact imaging of soft organic layers under sufficiently high loads. The scale of modifications is greater than 100 nm. The limits of direct surface manipulation were explored using the AFM. The system they used was a thin 50 A° metal oxide film made up of MoO_3 on the surface of MoS_2. The patent preached by Harvard University has some advantages over previous methods such as: (1) rigid and non-deformable thin MoO_3 films compared with organic layers; (2) MoO_3 can be machined or imaged depending on the applied load of the AFM cantilever; and (3) MoS_2 substrate acts as a stop layer. Patterns with less than 10-nm resolution can be achieved in this manner. Crystallites of α-MoO_3 were grown on the surface of single-crystal 2H-MoS_2 by thermal oxidation using purified oxygen at 480°C for 5 to 10 min. AFM images along with TEM and X-ray photoemission spectroscopy were used to characterize the nanostructures. The crystallites grow in the perpendicular axis to the substrate layer. Crystallites 1 to 3 unit cell thick and 200 to 500 nm on edges are formed.

The world's highest-resolution AFM as of December 2008 was developed by Asylum Research and was named Cypher. A closed-loop system with sensors in all three spatial dimensions was used in order to achieve atomic resolution. Features include automatic laser alignment and interchangeable light source modules with laser spot sizes as small as 3 μm and cantilevers smaller than 10 μm.

Rao et al.[20] explored a novel method for preparation of graphene containing two to four layers on a large scale. The method involves arc evaporation of graphite electrodes in a hydrogen atmosphere. The presence of hydrogen prevents the formation of closed structures. High currents >100 A, high voltages >50 V, and high pressures >200 Torr are favorable for graphene production. When hydrogen pressure is decreased, closed-shell polyhedral particles are formed. Diborane, pyridine, and ammonia are added in order to dope the graphene with boron and nitrogen. Reliable methods are needed to achieve graphene with high purity and doped samples with the desired number of layers. In Figure 2.6 an AFM image of graphene prepared by an arc discharge process from graphite is shown.

Hydrogen bonding is responsible for water being denser than ice at temperatures 0 to 4°K; speeding up or slowing down of reactions; causing the secondary structure of protein to be helix, sheet, or coil; and giving the DNA its shape and polymersomes their folded architecture and supramolecular structures. Hydrogen bonding has been largely explained by theory and very little direct experimental observations have been reported. Zhang et al.,[21] by using high-resolution AFM, have obtained spectacular images that can be used in elucidation of hydrogen bonds.

AFM has been used to show the atoms in bonds in a single pentacene molecule. The bond order of complex molecules has been revealed and photographs of molecules before and after undergoing chemical reaction have been developed from AFM images. Hydrogen bonds have weak attraction forces between atoms, and a hydrogen atom that is covalently bonded to an atom is more electronegative compared with hydrogen. Lone pairs of electrons are attracted partially toward the electropositive hydrogen. These bonds are not strong enough for the formation of covalent bonds. A team of scientists from National Center for Nanoscience and Technology in China found at temperatures near absolute zero that 8-hydroxyquinoline formed hydrogen-bonded aggregates. The electron

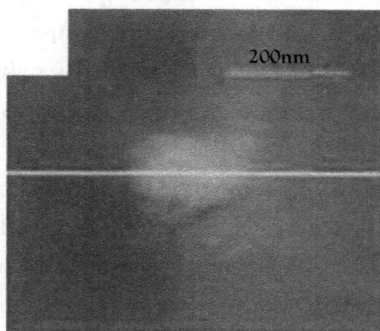

Figure 2.6 An AFM image of graphene made from graphite by arc discharge in an hydrogen atmosphere

densities of hydrogen bonds are made visible using AFM. They used non-contact AFM to observe intermolecular forces between a host of compounds. At room temperatures, hydrogen bonds were found in dimers and trimers of 8-hydroxyquinoline radicals complexed with copper. These can be formed by dehydrogenation of hydroxyl groups on copper surfaces.

At IBM's Zurich Research Center, seminal work on visualization of pentacene using AFM studies have been undertaken. Properties and mechanisms of hydrogen bonds can be better understood by combination of AFM and density functional theory. A chemistry professor at Indian Institute of Science, Bangalore, India, contends that hydrogen bonds have some covalent character. They have proposed a new definition of hydrogen bond at the International Union of Pure and Applied Chemistry (IUPAC) in 2010.

X-Ray Diffraction

Sir William Lawrence Bragg along with his father Sir William Bragg won the Nobel Prize for Physics in 1915 for their work on determination of crystal structure using principles of XRD. Lawrence Bragg is the youngest laureate as he received the prize when he was 25 years old. He was the discoverer of Bragg's law of XRD. He made it possible to calculate the positions of the atoms within a crystal from the way in which an X-ray beam is diffracted by the crystal lattice. He made this discovery during his first year of research at Cambridge and later he developed his ideas with this father and the X-ray spectrometer was developed.

Miller indices are used as the convention to specify directions and planes in a crystal.[22]

The crystal direction is denoted, for example, by [240], where 2, 4, and 0 are the components of a three-dimensional vector in the respective directions. Negative components are denoted by a bar. For example, the family [240], [420], [204], [402], and so forth, is denoted by <240>. The lattice parameter and Miller indices can be related to the interplanar spacing d as follows:

$$d = \frac{a}{\sqrt{h^2 + k^2 + l^2}}$$

Four-digit notation is used for hexagonal crystals. The *hkil* notation is called the Miller–Bravais indices. This notation is used to denote crystallographically equivalent planes or directions in a hexagonal crystal. Three of the axes, a_1, a_2 and a_3, are coplanar and lie on the basal plane of the hexagonal prism with an angle of 120° between them. The fourth axis is the *c*-axis orthogonal to the base plane. Two examples for determination of *hkil* can be seen in Figure 2.7. One of the vertical faces of the hexagonal prism is given by (1010). An intercept of 1 along the a_1 axis and an intercept of −1 along the a_3 axis are made and the plane is found parallel to the *c*-axis. In the Miller–Bravais notation, $h + k = -i$. The direction that can be seen lying along one of the three axes parallel to the side of the hexagon can be written as [2110].

In order for visible electromagnetic radiation to be diffracted, the spacing between lines in a two-dimensional grating must be of the same order as the wavelength of light ($\lambda \sim 3800$ to 7800 A°). This rule can be extended to the three-dimensional grating of the periodic array of atoms in crystals. The typical interatomic spacing in crystals is ~ 2 to 3 A°. The wavelength of the radiation used in crystal diffraction should lie in the same range. The wavelengths of X-rays lie in this range and are found to be diffracted by crystals. This property is used widely for determing crystal structures.

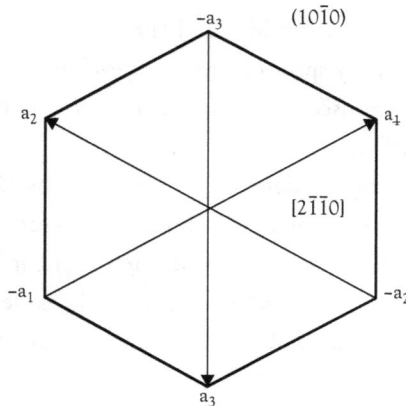

Figure 2.7 Planar view of a hexagonal unit cell. Plane (1010) and the direction [2110] in the Miller–Bravais notation

Electrons moving at high speeds when bombarded on a metal target lose part of their kinetic energies. This energy is converted to X-rays. White radiation refers to X-rays emitted from a target comprising a continuous range of wavelengths. The minimum wavelength in the continuous spectrum is inversely proportional to the applied voltage.

The acceleration of electrons is effected by the applied voltage. When the applied voltage is higher than a threshold value in addition to the white radiation, a characteristic radiation at a specific wavelength and high intensity is also emitted from the target. For example, both white and characteristic radiations are emitted from a molybdenum target at 35 kV. The characteristic radiations are named as K_α, K_β, L_α, and so on and so forth. The X-ray beam that is incident on the metal target interacts with electrons of the atoms in the crystal. Electrons oscillate upon the receipt of a new bout of energy. Waves emitted by electrons have the same frequency as the incident X-rays. Emission occurs in all directions. Emission in a particular direction is the result of combined effects of oscillations of electrons of all the atoms. The emissions can be expected to be in phase and reinforce one another only in certain specific directions. This depends on the direction of the incident X-rays, their wavelengths, and the spacing between atoms in the crystal. In other directions, destructive interference of the emissions from different sources occur. Bragg's law can be written as:

$$n\lambda = 2d\sin(\theta)$$

where n is the integer and λ is the wavelength of the X-rays used. n is used to denote the order of reflection. The maximum value of the sine function is 1. Hence, for an interplanar spacing of 1 A°, the Bragg's law can be used to determine the upper limit of the wavelength in order to obtain the first-order reflection as 2 A°. There will be no reflection if the wavelength is greater than 2 A°. Bragg's law can be used to obtain the lattice parameters of crystals. Reinforcing information can be obtained from higherorder reflections.

The powder method is widely used for determining crystal structures. The powder camera used is a Debye–Sherrer camera. This consists of a cylindrical cassette with a strip of a photographic film positioned around the periphery of the cassette. The powder specimen is placed at the center of the cassette made of a non-diffracting material. The X-ray beam incident through a small hole is allowed to pass through the specimen and the unused part of the beam is allowed to leave through the hole at the opposite end. The combined Miller–Bragg's law can be written as follows:

$$\sin^2(\theta) = \frac{\lambda^2}{4a^2}(h^2 + k^2 + l^2)$$

The θ values can be obtained from the powder pattern. Monochromatic radiation is used and the wavelength is known. Extinction rules for different crystal structures are drawn up. Hume Rothery rules can be used. For example, when h, k, and l are all odd or when all are even and $(h + k + l)$ divisible by 4, the crystal will be a diamond cubic (DC).

Kishore et al.[23] characterized graphene samples using XRD methods. In Figure 2.8, the XRD patterns of graphite, graphitic oxide (GO), com-graphene, and T250 graphene are shown. The main peak for graphite is ~26° corresponding to (002) reflection. This peak is not present in the GO sample. In GO, the main peak is at ~10°. Com-graphene was prepared by a combustion synthesis method. Urea was used as the fuel and ammonium nitrate was used as an oxidizer. One hundred milligrams of GO was dispersed in 50 mL of deionized water with ultrasonication for

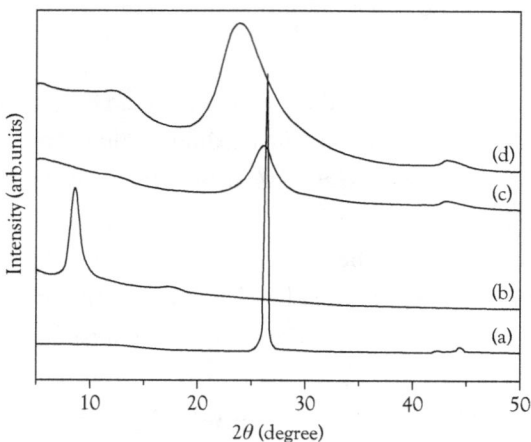

Figure 2.8 XRD patterns of: (a) graphite; (b) graphitic oxide; (c) com-graphene; and (d) T250 graphene

30 min. This was mixed with another solution prepared by dispersion of 0.9 gm of urea in 1.2 gm of ammonium nitrate dissolved in 5 mL of deionized (DI) water. The combined solution was stirred and heated to 150°C. A thick colloidal suspension was formed. The colloidal suspension was heated to 250°C for 2 h in air on a hot plate. The thermal treatment resulted in the removal of vapors, leaving behind a powder black in color. This product was washed several times with DI water followed by washing with methanol and dried at 60°C. This product is called com-graphene. T250 graphene is produced when the GO suspension is heated to 250°C for 2 h without fuel.

An XRD pattern of com-graphene shows a broad diffraction peak at ~26° and a peak from the precursor is absent. The functional groups are removed and graphene nanosheets are formed. T250 graphene is characterized by a peak centered at ~23° with a shoulder at ~25°. The interplanar spacing, d, is 3.723 A° for T250 and 3.370 A° for com-graphene. The crystallite thickness is ~35 nm corresponding to ~10 layers. Larger interplanar spacing would mean that functional groups are present between the layers. Large interplanar spacing can be used to detect SG.

Summary

As per the Raleigh criterion,[24] the resolution limit using optical devices is 200 nm—half the wavelength of light. X-ray devices are needed for observing nanostructures. In Table 2.1 the resolution size of different instruments is shown. A helium ion is the smallest ion available and an HeIM has greater resolution power compared with SPM, which in turn has a greater resolution power compared with TEM, which in turn has a greater resolution power compared with SEM, which in turn has a greater resolution power compared with optical microscopes.

Raman microscopes can be used to characterize single- and multiple-layer graphene samples (Figure 2.1). They can be used to measure temperature and crystal orientation of the sample. The G mode can be used to obtain the dopant characteristics. Intel has patented a Raman apparatus for detecting gene expression and genome sequence deduction.

Wavelengths associated with helium ions are shorter compared with even electrons. A world record on resolution was broken using an HeIM.

A surface resolution of 240 pm was achieved. It is three times better than that of SEMs with the same surface sensitivity. A new benchmark was achieved for surface imaging in the subnanometer range. The chemical composition of a defect in a semiconductor chip can be detected using an HeIM. Scientists at Harvard University have reported the etching of graphene devices using helium ion beams. They can also perform in situ electrical measurements during lithographic processing. Failures in electronics can be studied using HeIM by the resolution of whisker formation and growth. Images can be collected in two modes: the RBI and secondary electron modes.

The nanoscale structure of advanced material systems can be studied using *SAXS* devices in terms of relevant parameters. For example, the combined effects of graphene nanosheets and shear flow on the crystallization behavior of iPP were investigated using SAXS techniques.[25] SAXS can be used in order to obtain information about the shape, size, internal structure, crystallinity, and porosity of nanostructured samples. Structural information in the scale of *5 to 25 nm* can be characterized using SAXS. X-ray scattering has been used to study material structures that possess long-range order. The SAXS pattern provides information on the overall size and shape of these particles. The orientation and nanostructure of the sample can be obtained. Atoms and interatomic distances scatter toward large angles. The *WAXS* pattern obtained provides information on the phase state, crystal symmetry, and the molecular structure. True SWAXS is a feature of SAXsess that allows for simultaneous measurement continuously from small and wide angles upto 40° with a consistent high resolution and without changing the instrumental setup.

TEM can resolve structures in the nanoscale region. The sample specimens have to have thicknesses down to *few hundred nanometers* depending on the operating voltage of the instrument. *HRTEM* can achieve resolution sizes down to 1 A°. An HRTEM image of SG is shown in Figure 2.3. In a recent patent from MIT,[26] low-resolution cryo-TEM images of dipolar *oligopeptides* that self-assemble to form regular structures are discussed.

Using electron channeling contrast, M. Knoll in 1935 obtained the world's first SEM image of silicon steel. The *spatial resolution* of the SEM depends on the wavelength of the electrons and the electro-optical system that produces the scanning beam. *Raster scanning* using electrons are used in order to obtain SEM images of the sample surface. The surface *topography*, *composition*, and other properties can be obtained from the scan. An SEM image of nanowires is shown in Figure 2.4. A topographical map is generated on the atomic scale using an SPM. The surface features and characteristics of the specimen are obtained at magnifications higher than a billion. A *tiny probe* with a sharp tip is used and is brought in close proximity to within 1 nm of the specimen surface followed by raster scanning. STM, AFM, NFOM, and MFM are also discussed. An STM image of graphene formation by surface decomposition is shown in Figure 2.5. The STM evolved into the AFM. AFM is a special type of SPM discussed in the section "Scanning Probe Microscope." The resolution achieved using an AFM is a fraction of a nanometer. An AFM image of graphene made from graphite by arc discharge is shown in Figure 2.6. AFM can be used to image hydrogen bonds.

Microwave spectroscopy was developed using PAT through quantum dots. Rabi oscillations were found to occur. Electromagnetic field interactions with oscillating valence electrons are used in time-dependent measurements. A "Carbino" spectrometer was used to obtain the broadband response in the frequency range 0.1 to 16 GHz of CVD graphene at near -zero temperatures.

A scan using an *Auger Electron Microscope* identifies the elemental composition of the analyzed surface. Multiplex scan quantitates the atomic concentration of the elements identified in the survey scans. Detection limits are one-tenth of 1% of the atomic composition of the elements.

XRD has been used to obtain the Bravais lattice structures in materials science. Graphene can also be characterized using XRD. The peak found in graphite broadens in graphene. Larger interlayer spacing d estimates can be used to confirm graphene. The graphene formed from different process conditions can be distinguished using the XRD spectra.

CHAPTER 3

Applications

Chapter Objectives

- Ultrafast electronics
- Supercapacitors
- Desalination
- LEDs
- Thermal management
- Transparent electrodes
- Solar cells
- Batteries
- Anticorrosion coating
- Bionic materials

- Electromagnetic shielding
- Oil spills
- Surface modifications
- Superconductors
- Rapid DNA sequencing
- Magnetic sensors
- Rusnano
- Nanorobots
- Nanoscale thermometer
- Chemical modification

By 2018, the graphene market is expected to rise to $100 million. According to the BBC, the graphene market is expected to grow to $67 million by the year 2015 and to $675 million by the year 2020. The production value of graphene in 2010 was 28 tons and is projected to grow to 573 tons by 2017. The market for nanotechnology products will reach $3 trillion by the year 2015. Thirty-nine different nanostructuring methods were reviewed.[1] Identified in 2004, the number of patents on graphene is 7,351. Chinese entities have obtained 2,204 patents, US entities have acquired 1,754 patents, South Korean entities have been granted 1,160 patents, and UK entities have received 54 patents. Sixteen of the 54 patents in the United Kingdom are held by Manchester University.

Ultrafast Electronics

One or more atom layers in thickness, potentially graphene can be used to effect further speed increases higher than 30 PHz of microprocessor speeds. No material other than graphene has superior field emitter properties. Two-hundred times stronger than steel, as elastic/flexible as rubber, and harder than diamond, graphene is 13 times more conductive than copper. Electron mobility in graphene has been found to be 200,000 cm^2 V^{-1} s^{-1}. Graphene is the thinnest material ever created and weighs next to nothing. Graphene is a semimetal and can be used as barristors in the electronics industry. It can be expected and is found to conduct heat well. A host of different applications are expected for graphene in computing, energy, and medicine. These are of a wide range: higher-capacity electrodes, antireflection coatings in solar cells, carbon composites for lighter-weight BMWs, panel displays in wireless telephones and laptops, and thermal management. Graphene may be used in cutting-edge cancer treatments and in feather-weight HD televisions. Monolayer graphene is gapless. In order to have a tunable bandgap, another layer of graphene is needed.

Further advances in genome sequencing and proteome sequencing can be expected by using graphenes. Microarray analysis has led to the completion of the human genome project (HGP) ahead of schedule. The capability to study more genes per biochip is increasing at a rate much like Moore's law in electronics. Scientists in the Netherlands claim that they have found a method of rapidly sequencing deoxyribonucleic acid (DNA) and ribonucleic acid (RNA) strands. They pass these strands through the nanometer-sized sieves in graphene sheets. A voltage is applied across the sheet. Each of the nucleotide bases—adenine (A), guanine (G), cytosine (C), and thymine (T) for DNA, and uracil (U) in place of thymine in RNA—has a unique effect on the conductance of graphene as they pass one at an instance of time. Sequence distribution of DNA and RNA is deduced from changes in voltage. Computerization of the procedure and use of sensors with shorter response times can lead to more rapid sequencing of DNA and/ or RNA. In the case of proteins, 20 different amino acids can be classified as: (1) polar; (2) nonpolar; (3) acidic; and (4) basic. As each amino acid passes through the hexagonal sieve, the change in electrical characteristics can be used to detect the amino acid sequence distribution by calibration.

Scientists at Stanford University, Stanford, CA, have discovered the vast potential of using graphene oxide (GO) as nanocarriers for drug delivery. Nanorobots can be designed using nanostructured materials such as fullerenes and graphenes. GO as a nanocarrier in drug delivery has been used to increase cytotoxicity in treatment of cancer. GO can be used in gene therapy and to treat genetic disorders such as cystic fibrosis and Parkinson's disease.

Graphene can be used to effect further speed increases compared with 30 PHz. Thirty petahertz is theoretically achievable with the silicon chips with expected gate width limitations of 10 to 15 nm. With the thickness of graphene in the sub-angstrom region, the potential for increased speed in processors made of graphene is high. Prof. Philip Kim at Columbia University and coworkers have set a world record in electron and hole mobility. They clocked a mobility of 120,000 $cm^2 V^{-1} s^{-1}$ of electrons and holes in annealed single-layer graphene (SG) materials at 240°K. The carrier density was 2 trillion cm^{-2}. An intrinsic mobility limit of 200,000 $cm^2 V^{-1} s^{-1}$ was noted by Prof. M. Fuhrer at the University of Maryland. These findings can lead to the development of the field of ultrafast electronics. Low noise can be seen with graphene systems. Graphene can be used as a channel in field-effect transistors (FETs). SG can be placed in appropriate substrates, thus paving the way for narrower gate widths. Graphene transistors were made by IBM in 2008 at 26-GHz speeds.[2] Indium phosphide transistors have been used to clock 1 THz.[3] The electron mobility in graphene is six times higher than that found in copper. The 26-GHz speed record has been broken using a graphene transistor.[4] The cutoff frequency is 100 GHz. Further miniaturization is possible. These transistors can be used in microwave communication and imaging systems. Researchers at IBM's TJ Watson Research Center in New York made their FET by heating a wafer of silicon carbide (SiC) in order to create a surface layer of carbon atoms in the form of graphene. Source and drain electrodes in parallel were deposited later on the graphene. Channels of exposed graphene were left between the electrodes.

The next step is the deposition of the insulating layer on the exposed graphene without adversely affecting its electronic properties. A 10-nm layer of polyhydroxystyrene was used. Using this, a conventional oxide layer was deposited and then a metallic gate electrode was created. The gate length is about 240 nm. This can be potentially scaled down for further improvement in device performance. Graphene transistors have a higher cutoff frequency compared with the silicon metal oxide semiconductor field-effect transistor (MOSFET) devices with the same gate length. Graphene with zero bandgap between valence and conduction electrons needs to be made switchable before use in computers. At the University of Manchester, Manchester, UK, nanodiodes and transistors based on single-layered architecture have been developed. Speeds higher than 1.5 THz have been demonstrated. Applications envisaged include high-speed electronics, far-infrared THz detection and emission, ultra-high sensitive chemical sensors, and so on and so forth.

IBM has developed a new strategy for the fabrication of graphene-based transistors. Materials and methods compatible with those currently used in microelectronics are used. The graphene-based devices are smaller and can outperform the current devices used in the market. Graphene electrodes have been included in radiofrequency (RF) transistors and rapidly acting signal amplifiers in wireless communications. Graphene can be prepared in an efficient manner using vapor deposition methods. Attempts that have been made to deposit a graphene film on a layer of silicon dioxide (SiO_2) have resulted in deteriorated electronic performance. Y. Wu and Y. M. Lin of IBM's TJ Watson Research Center developed a vapor deposition method where graphene is atop diamond. This keeps the electronic performance high. When used in RF transistors, an increase in the frequency of operation was noticed. They can also work well at cryogenic temperatures. This approach is compatible with common methods used in the semiconductor industry. The methodology of fabrication has to be optimized.

As packing silicon chips reaches physical limits of miniaturization, alternative technologies for faster computers that are more energy efficient are explored. A research program at the University of California at Riverside[5] is dedicated to the development of computing using magnetic poles of electrons traveling in an SG sheet. Applications such as database searching, data compression, and image recognition where large amounts of data are involved will be accelerated using the developed electronic device called a "magnetologic gate" (MLG). This shall serve as a building block for circuits that can be used to combine functions of microprocessors and hard drives into a single chip. The research team comprises physicists, material scientists, and electrical engineers. The team has demonstrated efficient injection and transport of spin in graphene at room temperature and design of novel MLGs and circuits that can enable a computing platform based on electron spin to store and process information. High-speed, low-power information processing can result from the research program. Computer architecture with integrated logic and memory can lead to obviation of the von Neumann bottleneck. Three functional challenges are identified by the group: (1) better understanding and optimization of the spin injection and transport in graphene; (2) finding a high-speed and low-power method of switching ferromagnetic electrodes that are compatible with graphene; and (3) designing and implementing compound metal oxide semiconductor (CMOS)-compatible MLG circuits.

Tunable bandgaps in graphenes can be achieved using bilayer graphene. Scientists at the University of California, Berkeley, CA, are able to adjust the bandgap by changing the voltage and by placing two sheets of graphene on top of each other and putting the layers between the two electrical gates. Tunable bandgaps in graphenes can be used in development of novel nanophotonic and nanoelectronic devices. Electrical transport properties in devices can be manipulated using materials with tunable bandgaps.

Carbon-based electronic devices can be made from stable one-dimensional carbon chains of up to 16 atoms. This was made possible by researchers in China and Japan in the development of molecular nanowires. Starting with a flake of graphite and using transmission electron microscopy (TEM) to irradiate the flake with an electron beam, the scientists thinned the flake to realize a single layer of graphene. They punched two holes through the layer with high-intensity radiation leading to the formation of graphene nanoribbons between the openings. Cumulene in which carbon atoms are connected by double bonds and polyyne in which the single and triple bonds alternate between carbon atoms can also be realized in a chain form. The connection between chains and graphene was not stable. The life of the chains was as long as 100 s. The breakage is by detachment of the end of the chain from the sheet. The one-dimensional carbon chains can be a basic component of electronic devices.

In addition to graphene, another one-atom-thick material has been identified. It is boron nitride (BN) according to the researchers in Germany.[6] As will be discussed in Chapter 5, different methods can be used to separate the atom-thick sheets of graphene from graphite, sometimes readily. An atom-thick BN is another example of a thin free-standing crystal. Microscopy samples of hexagonal BN were prepared by peeling apart thin crystals of the sheet using an adhesive tape. An electron beam was used to thin the sample into one atomic layer. By tuning the beam intensity, the rate at which the beam ablated away the film was controlled. The sputtering process was analyzed by recording atomic resolution images and videos. These crystals were transferred to a microscopy grid. This study has improved the understanding of two-dimensional crystals in general.

Supercapacitors

Scientists at the University of Texas at Austin, TX, have found a way to store electrical charges in SG structures. These ultracapacitors can be used in renewable energy devices such as wind and solar power devices. Graphene can be used to double the capacity of current ultracapacitors. Ruoff[7] uses graphene sheets to rapidly store and discharge electric charge (EC). Rechargeable batteries and ultracapacitors are two devices of electrical charge storage. Ultracapacitors can be used directly or in conjunction with fuel cells and batteries as power sources. Chemically modified graphene was prepared. Its storage can be vital in enabling Sun Shot initiatives. This is so when large quantities of renewable electrical energy needs to be generated. Chemically modified graphene can be prepared in two steps of exfoliation of graphite powders and reduction of a graphite oxide product. A reduced GO (r-GO) product has attracted lot of attention in making ultracapacitor devices.

One possible problem in using graphene to prepare ultracapacitors with high power and energy density is the tendency of the two-dimensional graphene sheets to aggregate during the electrode fabrication process and align in the direction perpendicular to the flow of electrons and ions. Even when graphene materials with high surface area and low pore tortuosity are selected, the available surface area for energy storage is reduced by aggregation. This limits the scalability of the device. Stacking of two-dimensional graphene layers leads to formation of lamellar microstructures parallel to the current collectors. This happens more so during the compression step needed to make the electrode. The adverse effects of this structure formation is twofold: (1) effective surface area of electrodes is reduced and (2) horizontal alignment of graphene stacks can obtrude electron and ion transport.

The aggregation is because of van der Waals forces of attraction. The force is proportional to the overlapping surface area (S) and fourth power of the inverse of separation distance between sheets. A method to overcome this difficulty was proposed by Luo et al.,[8] to transform the two-dimensional graphene sheet into a crumbled paper-ball structure. Crumbled balls can be used to deliver higher specific capacitances compared with flat or wrinkled sheets. The method used to prepare crumbled GO balls was isotropic capillary compression in rapidly evaporating droplets of GO dispersed using an aerosol spray pyrolysis setup. Collected fine powders of the product were redispersed in water and reduced using N_2H_4. They[9] found that conversion of flat graphene sheets into crumbled paper-ball morphology can significantly improve the scalability of graphene-based ultracapacitors. The aggregation-resistance properties of the crumbled particles allow for retention of the high accessible surface area and high specific capacitance. The void spaces within the crumbled paper structure can be used for charge transport at high current density and high mass loading. This can lead to high specific capacitance. The capability is rated by mass and volume. Use of crumbled structures in place of sheet structures can obviate the need for binder materials such as polytetrafluoroethylene (PTFE) colloids, leading to further improvement in the performance of the device. Crumbled-ball morphology can be used in preparation of nanocomposites,

Nanostructured spacer materials can be added to prevent stacking of graphene sheets. Paper balls of graphene are aggregation resistant. They are dispersible in common solvents. The strategy may work as soccer-ball structures of carbon, C_{60}, have been found to work well in nanocomposites. The crumbled-ball structure can be used to outperform the flat sheet structure in catalysts, ultracapacitors, and microbial fuel cells.

Desalination

Sea water contains about 3.6 wt% sodium chloride salt. The current methods used for obtaining potable water from sea water are reverse osmosis (RO) and electrodialysis. RO differs from the ultrafiltration method in the size of solute that is rejected. Two types of RO membranes are used: (1) asymmetric membranes and (2) thin-film composite polymeric membranes. A chart that contains water flux per unit interfacial area per day versus salt rejection rate for RO processes is presented in the work by Sharma.[10] Data from 19 different commercially available membranes were included in the chart. The equations that can be used to describe the RO processes were developed using the principles of osmotic pressure, permeability of solvent across the membrane, and diffusion of solutes across the membranes. Osmosis is the flow of a solvent from a region of low solute concentration to a region of higher solute concentration. An expression for the osmotic pressure across a hydrophilic membrane was developed from equilibrium thermodynamics and concept of fugacity. This relationship is the van't Hoff's law. The use of Taylor approximation can be used to make calculations simpler in different applications.

The effect of solute concentration on the permeate side was accounted for and the logarithmic functionality was retained in the extended van't Hoff's law. The combined effect of osmotic pressure and hydraulic permeability was accounted for using Starling's law. The hydraulic conductance in terms of the characteristics of the pores in the membrane was presented. The sieve coefficient developed by Deen can be used to obtain the solute concentration in the permeate side. The solute diffusion across the membrane can be accounted for using Fick's laws of diffusion and generalized Fick's laws of diffusion. The solute rejection can be estimated using the Staverman reflection coefficient. The pore radius information of the membrane and other characteristics are needed. Recycling can result in significant cost saving at the RO pump.

Ninety-seven percent of the world's water resources lies in the oceans. Only a fraction of a percent of the world's potable water supply is obtained from the seas. The RO processes are used in areas where the energy is cheap and water is scarce such as in the deserts of Sahara or the deserts of Nevada. An energy efficiency of 1.8 kWh m^{-3} was achieved at a commercial desalination plant recently. This is an improvement from ~5 kWh m^{-3} in the 1990s.

The water flux across the RO membranes scales inversely with the membrane thickness. Two-dimensional graphene sheets offer tremendous potential in this area. The advantages of using two-dimensional graphene over convention RO membranes are negligible thickness, high mechanical strength, low pressure drop requirement for the RO pump, and wide range of operating conditions. Nanopores can be introduced into graphene's structure. Cohen-Tanugi and Grossman[11] used molecular dynamics simulations and examined the desalination dynamics using SG materials. They found that water can flow across the graphene membrane at rates of 10 to 100 L cm^{-2} day^{-1} MPa^{-1}, with salt rejection of 2 to 3 orders of magnitude higher than that found in RO membranes where the mechanism is largely diffusion. The pore sizes examined in the study were in the range of 1.5 to 62 A°. The study included pores passivated by hydrophilic groups as well as hydrophobic groups. Water was modeled using the TIP4P potential and interactions of all other atomic species were modeled using Lennard-Jones (LJ), Coulombic terms. Simulations were performed using the Large Scale Atomic/Molecular Massively Parallel Simulator (LAMMPS) package.

Their molecular dynamics (MD) simulations indicate that nanoporous graphene membranes can be used to reject salt ions while allowing for water flow at permeabilities several orders of magnitude higher than those of existing RO membranes. Desalination performance was found to be sensitive to pore size and pore chemistry. Examination of water structure near the pores revealed that the hydrophobic character of the hydrogenated pores can result in reduction of water flow by imposition of additional conformation order. Limited hydrogen bonding allows for greater salt rejection relative to hydroxylated pores.

Hydroxylated and hydrogenated graphene can result in improvements in water flux from current 0.01 to 0.1 L cm^{-2} day^{-1} MPa^{-1} to nearly 100 L cm^{-2} day^{-1} MPa^{-1}. The MD study can be used in redesigning of desalination membrane materials using SG materials.

Capacitive deionization (CDI) is a technology that has been developed in order to remove charged ionic species from salty water. Electrosorption is a method in which charged ions are forced toward oppositely charged electrodes by imposition of a direct electric field in order to form a strong electrical double layer and hold the ions. CDI can be a lower-energy-consumption route for desalination. Graphene nanoflakes (GNFs) can be used as electrodes for CDI. With its high interfacial area, the electrosorption performance of GNFs was much higher than that of commercial activated carbon. Electrosorption isotherms and kinetics were investigated.[12] The specific surface area of GNFs was found to be 222.01 m^2 g^{-1} and their specific electrosorptive capacity was 23.2 μmol g^{-1} for Na$^+$ ions when the initial concentration (C_{A0}) was 25 mg L^{-1}. The Langmuir isotherm was confirmed. Rate constant, equilibrium, and electrosorption capacity were reported. Once the electric field is switched off, the ions that were held are instantly released back to the bulk solution. CDI can potentially have lower energy consumption, easy regeneration of electrodes, and is environmentally friendly. Salt ions are removed using this technology and not pure water as in other technologies such as RO, and so forth. Graphene possesses a large specific area of about 2,600 m^2 g^{-1}, a tensile modulus of up to 35 GPa, and a higher room temperature electrical conductivity of 7,200 siemens m^{-1}. These properties can lead to an interesting graphene electrode. Electrosorption of sodium chloride using graphene electrodes was found to be 20.1 μmol g^{-1} and a lower specific surface area of 14.2 m^2 g^{-1} was observed. Further increases in electrosorption capacity can be found by increasing the surface area of graphene. The key factors that affect the electrosorptive performance of Na$^+$ ions on the surface of GNFs are the flow rate and bias potential. Higher voltage leads to higher EC due to stronger Coulombic interactions. A fivefold increase in electrosorptive capacity from 4.44 to 23.2 μmol g^{-1} was found when the voltage was increased from 0.8 to 2.0 V. A maximum in the electrosorptive capacity of GNFs was found when the flow rate was increased from 15 to 25 mL min^{-1}. Further increases in flow rate resulted in decrease in electrosorptive capacity. The occurrence of the maximum is attributable to equilibrium between electrostatic forces and hydrostatic flow forces. The initial concentration of Na$^+$ ions was varied and the Langmuir kinetics was confirmed from the absorption study. Monolayer adsorption was found to be dominant during the electrosorption process. The adsorption rate constants of Na$^+$ ions were determined using the Lagrangian equation:

$$Log(q_e - q) - \log(q_e) - 0.4342k_t \qquad (3.1)$$

where k is the adsorption rate constant (min^{-1}) and q_e and q are the amounts of Na$^+$ ions adsorbed at equilibrium and at time t (min) respectively. GNFs were found to have better performance as electrode materials. Optimal working conditions and bias potential were identified. GNFs have been found to have high potential in brackish water desalination and drinking water purification.

Light-Emitting Diodes

The operation of light-emitting diodes (LEDs) is based on electro-photoluminescence phenomena. Light is emitted from a semiconductor material when current or voltage is applied. LEDs prepared using p-type graphene doped with a p-type dopant and n-type graphene doped with an n-type dopant are patented by Samsung LEDs Co.[13] A graphene superlattice with graphene nanoribbons is used to connect the p-type and n-type graphene. The graphene nanoribbon is zig-zag shaped on opposite edges. The width of the nanoribbon is 3 to 20 nm. The interval between adjacent nanoribbons is 2 to 15 nm. The graphene superlattice has periodically arranged quantum dot patterns. The quantum dots can have the same size/shape or may be allowed to differ in size/shape. Depending on the size, the quantum dots may correspond to red, green, and blue photoluminescence. p-Type dopants may be oxygen, gold, or bismuth atoms or acidic compounds. The concentration level is about 10^{-20} to 10^{-5} cm^{-2}. n-Type dopants can be nitrogen, fluorine, manganese atoms, or ammonia. The concentration of the dopant is between 10^{-20} and 10^{-5} cm^{-2}. The p-type and n-type dopants may be spaced apart at constant distances. When voltage is applied to the p-type graphene and n-type graphene in the forward direction, holes in the p-type graphene and electrons in the n-type graphene move toward the active graphene. Energy bandgap with the dopant can be seen for both p-type and n-type graphene. Electrons and holes recombine in an active graphene superlattice and emit photons. Light corresponding to the bandgap is emitted. The higher electron mobility in graphenes translated into high brightness emission.

A graphene FET is shown in Figure 3.1 that is used in the study of electrical characteristics of a p-type and n-type graphene patented by Samsung LEDs.[14] The current–voltage characteristics (I vs. V) of the photoluminescent device can be seen to be nonlinear and deviating from the predictions of Ohm's law of electricity. Dirac voltage at low currents and a concave curvature in the curve in the electron flow region (p doping) can be seen. The curve is a mirror reflection of the electron flow region in the hole flow (n doping) region of the graph. An enlarged view of the quantum dot can be seen in Figure 3.2. Graphene quantum dots can have the same diameter and may be dispersed between p-type graphene and n-type graphene. The size range of quantum dots is 1 to 100 nm.

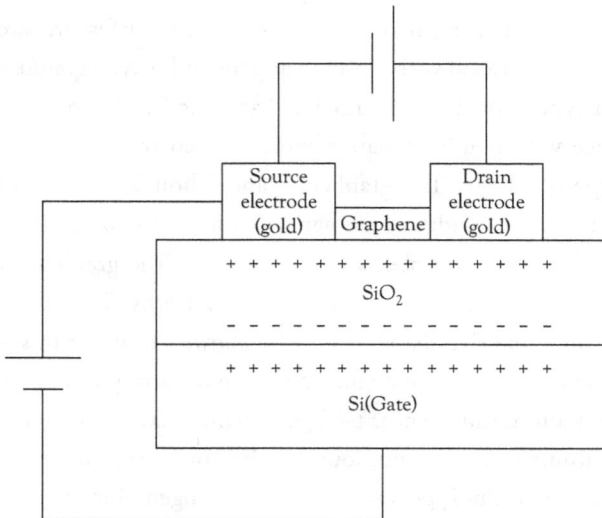

Figure 3.1 Study of electrical characteristics in a light-emitting graphene FET

The size of the quantum dots can be directly related to the photoluminescence wavelength. Longer photoluminescence wavelengths can be seen from larger quantum dots. The photoluminescence characteristics such as mono color or dichromatic or multicolor can be controlled by controlling the size of the quantum dots. When the LED is made from compound semiconductors, the composition of the compound can be changed in order to control the output photoluminescence characteristics. By attaching a functional group to the quantum dot, the bandgap or doping characteristics of the LED can be changed. When a voltage is applied to the p-type and n-type graphene, holes and electrons move toward the active graphene. Electrons and holes recombine and a photon is emitted corresponding to the energy bandgap. Photoluminescence caused by quantum dots of the active graphene may result in amplification of light generated due to the recombination of holes and electrons. Due to the higher mobility in graphene, a higher level of current may be supplied in order to achieve high brightness emissions.

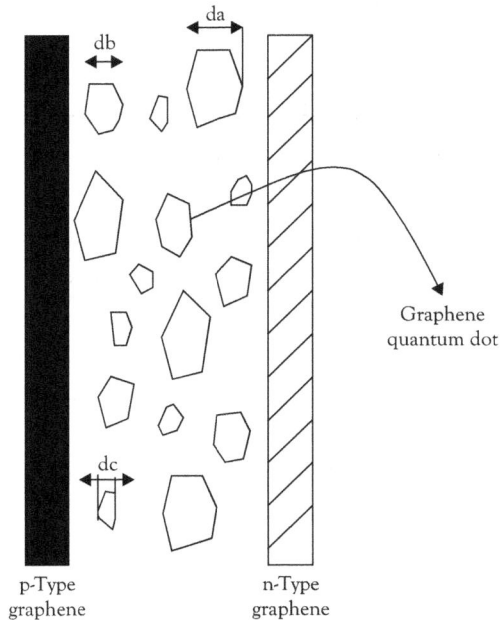

Figure 3.2 Graphene quantum dots in the Samsung patent

Thermal Management

The Boeing Company has patented a metal matrix composite (MMC) with graphene nanoplatelets dispersed in a metal matrix.[15] With enhanced thermal conductivity, these MMCs can be used to make heat spreaders and other thermal management devices. Improved cooling can be provided to electronic and electrical equipment and semiconductor devices. The graphene nanoplatelets have thicknesses less than 100 nm. The graphene layers are made with orientation with the matrix material. The metal matrix is densified by the use of graphenes. Semiconductor devices are mounted on the MMC.

Materials with high thermal conductivity can be used to form heat spreaders and heat sinks in high-power electronic packaging and other thermal management applications. Material properties can be optimized by addition of fillers to metal matrices. Coefficient of thermal expansion (CTE), wear resistance, and stiffness are properties of materials that can be improved by use of MMCs. One disadvantage of using MMCs is the poor stress characteristics. The density of graphene is approximately 1.7 g cm^{-3}. The height of the graphene sheets can be less than 1 nm.

The graphene layers can be made to have uniform orientation in the matrix. The metal matrix can be made out of aluminium and copper. The composite can be made with different volume fractions of graphene. The thermal conductivity of the graphene nanoplatelets is about 5,300 W m^{-1} K^{-1}. The thermal conductivity of the MMC would be higher in the direction parallel to the graphene sheet layer and lower in the direction perpendicular to the graphene sheet layer.

A thermal conductivity of 713 W m^{-1} K^{-1} in the plane of a graphene sheet was reported. The thermal conductivity to weight ratio of the MMC was higher than that of aluminium. When copper was used, the values of thermal conductivity went up to 860 W m^{-1} K^{-1}. The MMC with three atomic layers of graphene sheets was tested in thermal management of electronic packages.

Graphene layers can be used to cool hotspots in processors. A heat-dissipating effect on silicon electronics has been found by a researcher Prof. J. Liu at the Chalmers University of Technology. The working temperature in hotspots found inside the microprocessor can be reduced by 25%.

This can lead to an increase in the service life of electronic devices. A rule of thumb in thermal management is that a 10°C rise in working temperature can result in a reduction by half in the service life of the electronic system. The hotspots studied are at a nanoscale. The range of temperatures of hotspots the researchers have worked with lies between 55 and 115°C. A reduction by 13°C in hotspot temperature was demonstrated by Liu.[16]

Efficient heat dissipation is a technical hurdle in a number of other applications such as automotive electronics, power electronics, computers, radio base stations, and LEDs. Ignition systems in automotive electronics can pump up to 80 W at steady state and 300 W in a transient period of about 10 ns. Thermal intensity on par with the sun, ~600 W cm^{-2}, can be found in LED devices. About 50% of total electricity used to run data servers goes into cooling these systems.

Transparent Electrodes

Graphene is an ideal choice for preparation of transparent conducting electrodes. Its atomic layer thickness, transparency, large surface area, inertness of oxygen, and high electron and high hole mobility can be tapped into. Graphene sheets have been found to absorb only 2.3% of white light. Eighty percent of transmittance from the graphene electrodes has been reported for materials grown on a 300-nm-thick layer corresponding to 6 to 10 graphene layers. Transmittance was found to increase to 93% by making the graphene film thinner. Indium tin oxide (ITO) is currently used in order to make transparent conductive coatings for liquid crystal displays (LCDs). A review[17] was presented on applications of graphene. Graphene-based electrodes for dye-sensitized solar cells were reported. The characteristics of the cells showed short-circuit photocurrent density of 1.01 mA cm^{-2}, open-circuit voltage of 7 V, calculated filling factor of 0.36, and overall power efficiency of 0.26%. Graphenes have been used in composites with poly(3,4-ethylenedioxythiopene) (PEDOT-polystyrenesulfonic acid [PSS]) and made into counter electrodes in order to increase the transmittance to greater than 80%.

Organic solar cells can also be made from graphene films. Power conversion efficiencies comparable with ITO-based solar cells were achieved using graphene sheets. Graphene sheets synthesized on nickel substrates were transferred to conductive substrates and used in transparent electrode applications. This was found suitable for making smaller electrodes. Research is underway to perform large-area graphene synthesis directly on transparent substrates. Flexible and stretchable electrodes can be made from transparent graphene films. Foldability of the graphene films was evaluated. Graphene is used as a novel acceptor for bulk heterojunction polymeric photovoltaic cells.

Solar Cells

The use of graphene layer in solar cells can lead to higher light to electricity conversion as much as 15% and also lead to lower costs. One reason for the higher costs of solar cells is the thin charge-carrying silicon layer on top. Processes developed to meet these specifications need operations at high temperatures. Thin silver wires serve as an electrode. A recent discovery in Peking University, China, showed that graphene films can replace these charge-carrying layers.[18] They can also be used as electrodes. This is less expensive to fabricate. But the light to electricity conversion efficiency is about 8.6 to 10.0%. The scientists at the National Center for Nanoscience and Technology at Beijing, China, coated a 65-nm thick layer of TiO_2 atop a silicon graphene solar cell. The amount of visible light reflected from the silicon surface is decreased to less than 10% from 30% on account of the coating. Less reflected light would mean more light to electricity conversion efficiency. The efficiency was found to increase to 14.6%. Graphene layers are grown on copper foils using chemical vapor deposition (CVD) and transferred to silicon wafers.

A carbon nanotube (CNT) in the unscrolled form is graphene. An epitaxial substrate is a prerequisite for preparing two-dimensional nanosheets. Atomic bonding in the third dimension (Landau–Peierls argument) is provided for by the substrate. This is a salient consideration when attempting to synthesize a material with thickness of an atom that is stable. The monolayer structure of graphene is confirmed using Raman spectra.

Another important stability consideration is the thermodynamic metastability. Gibbs free energy change of formation of graphene structures has to be negative for the formed structures to be stable.

Unstable structures may be expected when the second derivative of the Gibbs free energy with respect to phase volume is negative. Metastable structures can result when the Gibbs free energy change is negative and the phase stability criterion is not met. Graphene sheets with less than 24,000 atoms/25 nm are metastable. The structure of graphenes has been confirmed as the honeycomb lattice structure.

Graphene made by CVD methods can be rolled into thin films by transfer processes. A typical process comprises three steps: (1) adhesion of polymer supports to the graphene on a copper foil; two rollers are used to get the graphene film grown on a copper foil, as it passes through to be attached to a polymer film coated with an adhesive film; (2) etching of copper layers. Graphene is synthesized in an annular plug flow reactor (APFR). The annular reactor space is generated by an 8-inch outer quartz tube and an inner 7.5-inch copper foil wrapped quartz tube. The use of an annular reactor in place of the tubular reactor is to minimize radial temperature gradients. This was found to cause inhomogeneity in the film formation. The inner tube is heated to 1000°C. Hydrogen (H_2) is allowed to flow at 8 sccm (standard cubic centimeters per minute) and 90 mTorr. The annealing process comes next. Annealment for 30 min allows for increase in grain size in copper foils from a few micrometers to 100 μm. This has been found to increase the graphene growth. Methane (CH_4) is allowed to mix with the flowing hydrogen at a flow rate of 24 sccm at 460 mTorr. (3) The sample is rapidly cooled at about 10°C s^{-1}. The graphene film grown on a copper foil is attached to a thermal release tape by applying pressure on the rollers at 0.2 MPa. The copper foil is etched in a plastic bath filled with an etchant. The etched film is washed with deionized water to remove any unused etchant. The graphene film is ready for transfer to a target substrate such as a curved surface. One hundred and fifty to two hundred millimeter per minute transfer rates by thermal treatment can be achieved by letting the graphene film pass through the rollers at a mild heat of 90 to 120°C. Multilayered graphene can be made by repetition of this process. The product that comes as a result of this would be different from the bilayer or multilayer material formed during the reaction by other methods. This can be viewed as physical stacking of the formed graphene layers. Screen printing can be used to generate four wire touch panels.

Continuous production of graphene on large scale is within reach. Scalability of the process is high. Processability is good. Carbon has limited solubility in copper at 1,000°C. The copper may have a catalytic effect on the graphene formation reaction. Monolayer graphene was confirmed using Raman spectra. Bilayer and multilayer islands were found using an atomic force microscope (AFM) and a TEM. Stacked layers reduce the optical transmittance by 2.2 to 2.3% a layer and the conductivity also decreases. Dopants can be added as desired. p-Junction formation by doping can be achieved by addition of nitric acid (HNO_3). Sheet resistance can be increased by chemical doping. Polymethylmethacrylate (PMMA) can be used as a polymer support. Some of the challenges in using this method is the formation of polycrystalline graphene due to the occurrence of nucleation again to form a second layer. Oxidation of copper has to be avoided. High rates of evaporation of copper from the foil can hinder graphene growth. Copper is not as effective as nickel in lowering the energy barrier to form graphene.

Further reductions in fabrication costs for SGs are achieved by painting or spraying a graphene solution onto the silicon wafers in order to produce the graphene film. These devices have to be made more stable. Solar device efficiency slips after 20 days of outdoor performance.

The reason why small differences in structures of materials can strongly affect performance needs to be better understood. Determining structure–property relationship in electrode materials is a technical challenge when nanoparticle aggregates are used. Scientists at the University of North Carolina have found that the relative orientation of adjacent nanoparticles significantly affects the charge transport. They used TEM (Chapter 2) in order to obtain images of the spatial distribution and orientation of nanocrystals within aggregates. They also used a conducting AFM method (Chapter 2) and obtained a correlation between the TEM information and the pathways that electrons follow during the passage through the material. Various nanoparticle-based iron electrodes were studied. Hydrogen is generated via light-driven water splitting. Some electrodes were found to work well while others did not. A small orientation mismatch among the nanoparticles is tolerable but a larger mismatch was found to block current flow between adjacent grains. Certain crystal orientations were called "winning" orientations and the process is geared toward enabling these orientations to come about. A world record in photocurrent was achieved. Scientists in Berlin, Germany, hailed this as an important step in the development of next-generation energy conversion devices.

Batteries

There is an increased demand for batteries for hand-held devices such as cell phones, iPads, and so forth. Lithium-ion batteries are a key component in these hand-held devices. Graphite is used as an anode material. Graphene-based anodes have been proposed as alternative materials. Graphene has superior electrical conductivity properties compared with graphitic carbon. It also possesses higher surface area; better chemical tolerance, energy density, durability, reversibility; and higher specific capacity. In their work, Choi et al. discussed a graphene nanosheet that is decorated with SnO_2 particles with a reversible capacity of 810 mAh g^{-1}. The nanosheet–SnO_2 structure was obtained by dispersion and reduction of graphene nanosheets in ethylene glycol and reassembly in the presence of SnO_2 nanoparticles (Figure 3.3). Its cycling performance is improved significantly by the use of SnO_2 particles alone. Self-assembled TiO_2–graphene nanostructures with high rate performance of the electrochemically active material can be another example. This structure is prepared using sulfate surfactants that help with the stabilization of graphene in aqueous solutions. The structure was found to have a specific capacity of 87 mAh g^{-1}.

At the 241st American Chemical Society National Meeting held at Anaheim, CA, in March 2011, a prototype version of a novel lithium-ion battery was described. These batteries are three-dimensional and can be used to recharge batteries in electric vehicles in minutes rather than hours. These batteries are expected to last longer. They can be used with other electronic devices as well. Dr. Amy Prieto has cofounded her own company Preito Battery. In another 2 years, three-dimensional batteries are expected to be available to the consumers. Laboratory tests confirmed that the batteries could be recharged in 12 min compared with 2 h for a standard lithium-ion battery. The service life of the batteries is doubled. More lithium per unit volume is achievable with the new technology.

Increased capacity: Li* can be intercalated into both GNS and SnO$_2$

Graphene nanosheets (GNS)   SnO$_2$/GNS

Enhanced cyclability via 3–D flexible structure

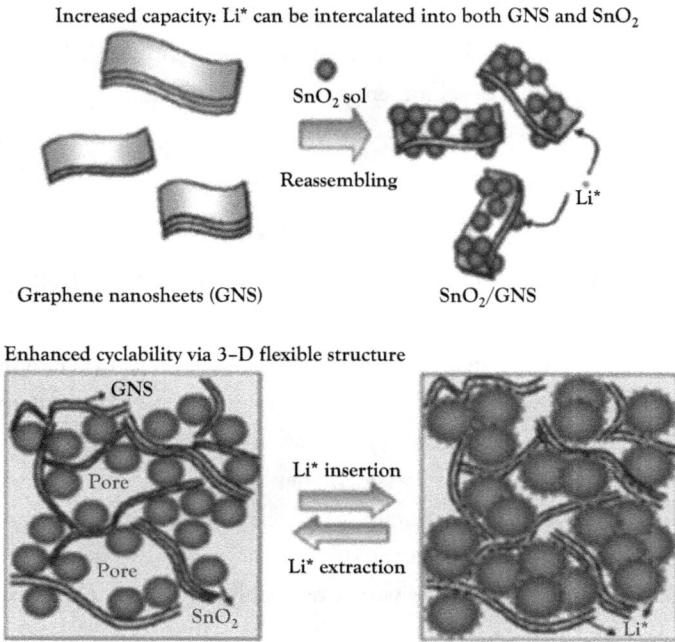

Figure 3.3 Intercalation of lithium ions on graphene nanosheets and SnO$_2$.

Source: Reprinted with permission from the American Chemical Society.

The size of the three-dimensional battery is about the size of a cell phone. Nanowires are arranged inside the battery into a three-dimensional structure with the appearance of a hair brush. The nanowires are coated with a thin layer of the electrolyte. This also forms the separation between the anode and the cathode. Graphene is used in place of graphite. They are also researching use of nanowires of copper and antimony.

Anticorrosion Coating

Graphene has been used to make the thinnest coating known world over and can be used for protecting metals against corrosion. The potential use of graphene as anticorrosion coating is discussed in *ACS Nano* by D. Prasai and his colleagues. Rusting and other corrosion of metals is an important global problem. Contact of metal with air, water, or other substances can cause corrosion. Graphene is a single layer of carbon atoms. It is evaluated for use as an anticorrosion coating. An ounce of graphene arranged in a single layer and comprised of rows of benzene rings can fill the size of 28 football fields. Graphene, whether made directly on copper or nickel or transferred onto another metal, can be used to prevent corrosion. Copper coated with graphene by CVD was found to corrode seven times faster than uncoated copper. Nickel coated with multiple layers of graphene was found to corrode at a rate 20 times slower than uncoated nickel. The same amount of corrosion protection can be obtained using a single layer of graphene that was obtained using five layers of organic coatings. Graphene coatings may be ideal for applications in industrial microelectronics. They can be used in aircrafts, implants, and as interconnects.

Bionic Materials

Graphene materials may be used to make bionic eyes and ears. The properties of graphene that makes it suitable for this application are as follows: (1) it is impervious to harsh ionic solutions and (2) electric signals can be conducted and can interface with neurons and other cells. In Schmidt's work,[20] a method is discussed to detect action potentials in heart cells using an array of graphene transistors. This is a big bionic step according to nanotechnologists at the Walter Schottky Institute in Munich, Germany. Neurocare is an European project with a budget of $6.3 million where carbon-based implants in ears, eyes, and the brain are evaluated. In addition to the chemical stability and electronic attributes, graphene can also be readily wrapped around delicate tissues. A number of research opportunities have arisen on account of the capability of graphene in the area of neural prosthetics. CVD was used in order to grow a layer of cardiomyocytes directly atop the graphene transistor array. Action potentials are used to pass electrical signals from cell to cell. This is common in electrogenic cells such as cardiomyocytes. The voltage fluctuations cause changes in the electrostatic environment in the electrolyte separating the cells and transistors. This induces a flow of ions in the electrolyte. The electrical resistance is altered in the transistors on account of the induced flow of ions. This constitutes a detectable electrical signal between the electrogenic cells. In order to make graphene devices suitable for digitization, they need to be switchable. Graphene transistors are used as nanosensors. Silicon transistors can be made more stable in solution by coating them with metal oxide. However, this coating will entrap ions that causes noise in the signals. Graphene transistors without any coating can be used with less noise.

The thickness of graphene can be used to improve the interface between retinal implants and eye tissues. Research is underway to stack graphene transistors on polyimide Parylene that tend to be biocompatible and flexible.

Infrared Transparent Electromagnetic Shielding

Graphene is an interesting material of choice in naval applications. It can be used to demonstrate an infrared-transparent, electromagnetic shielding coating that can be applied to electro-optic sensor windows and domes. Transmission greater than 90% in the 3 to 5 μm wavelength region can be achieved. Sheet resistance can be made less than 10 Ω per square. The coating can be found to be chemically stable in the atmosphere and in daylight and exhibit good resistance to rain and solid particles. The coating temperature used can be as high as 600°C.

Electromagnetic shielding of electro-optic sensor electronics in military systems is currently provided by metal grids that are electrically conductive applied to the dome or sensor window. However, the optical system is compromised via geometric blockage, diffraction, and unwanted reflection of light. Grids provide electrical shielding but are difficult to be deposited on curved shapes. Damage is caused by erosion from rain and particle impact on the external surface of the dome or window. Graphene thin-film coating with high electrical conductivity and optical transparency can provide adequate electromagnetic shielding and superior optical performance. Owing to its hardness, erosion resistance can be expected to be high. The graphene layer can be part of a stack of layers in order to minimize reflection losses and provide high erosion resistance. The coating can be deposited on an infrared-transparent substrate such as sapphire or silicon. This allows for infrared optical properties to be measured. Silicon can be used only if it is chemically inert under the film deposition conditions.

White Graphene and Oil Spills

Scientists have made nanosheets of BN also called as "white graphene." These nanosheets have 33 times greater absorption potential than certain chemicals. These sheets can be used in abatement of water pollution. Spilled oils can be soaked up and chemical solvents and dyes discharged by leather and textile industries can be absorbed using "white graphene." These sheets have a high porosity and possess high surface area. They can float on water and are hydrophobic. A team from France and Australia said that once the white sheets are dropped on an oil-polluted water surface, the brown oil is immediately absorbed and a color change to dark brown can be observed. This absorption process takes about 2 min. Upon reaching a state of saturation, the sheets can be cleaned by washing, heating, or combustion. The sheets can be reused. The nanosheets have a much higher absorption potential compared with carbon or natural fibers that are currently used in oil spill abatement.

Surface Modifications on Graphene

In order to fabricate electronic devices that tap into the better electrical conductivity and strength properties of graphene, the graphene needs to be chemically modified. Sensors can be made by attachment of biomolecules onto graphene. Patterning graphene sheets with impurities can make them semiconductors. A. B. Braunschweig at the University of Miami and K. N. Houk at the University of California at Los Angeles[21] have demonstrated that pressure can be used to accelerate production of patterns of covalent modifications via Diels–Alder adduct formation on a graphene surface. A 2 × 3 array of 80-nm-wide polymer tips was coated with inks containing various molecules linked to cyclopentadiene. For one of the inks, the dye cyanine 3 was attached to cyclopentadiene. The tip array was mounted on an AFM and the tips were gently pushed onto the surface of graphene. Graphene and cyclopentadiene reactions were facilitated resulting in 20 μm × 40 μm patches of graphene decorated with a 2 × 3 pattern of dye dots. The technique is performed at room temperatures and pressures. Attempts are underway to have graphene functionalized with carbohydrates in order to make biological sensors.

Graphene Nanocomposites

Toyota commercialized clay nanocomposites made of nylon-6 and 5% clay for timing belt covers. General Motors (GM) had its first exterior trim application of nanocomposites in their 2002 mid-sized vans. The part was stiffer, lighter, and less brittle at cold temperatures and more recyclable. GM uses about half a million pounds of nanocomposites each year. By 2003, the worldwide consumption of polymer nanocomposites was valued close to $100 million.[22] The word nanocomposite was coined by Komerneni. The dispersed phase comprised 10-nm regions in ceramic metal composites. Solids with more than one Gibbsian phase with atleast one dimension in the nanoscale region may be called as nanocomposites. The state of a material can be crystalline, semicrystalline, or amorphous. The morphologies can be immiscible, exfoliated, or intercalated.

Polymer nanocomposites are emerging as a topic in the materials science curriculum. Examples are development of exfoliated clay nanocomposites, CNTs, exfoliated graphite, nanocrystalline metals and other filler-modified composite materials, exfoliated graphene, fullerene nanocomposites, and so forth. Performance enhancement includes increased barrier properties, flammability resistance, electronic properties, membrane preparation, polymer blend compatibilization, and so forth. Clay platelets are 9.4 A° in thickness. Wide-angle X-ray scattering (WAXS) spectra are used to relate the morphological structure to the performance properties of nanocomposites.

Qi et al.[23] compared the electrical conductivity of multiwalled CNT (MWCNT)–polystyrene (PS) nanocomposites with that of graphene–PS composites. They recorded a 2 to 4 orders of magnitude increase in electrical conductivity when MWCNTs were replaced with graphene. The graphene volume content was 0.11 to 1.1%. The electrical conductivity values of the graphene composite was ~3.5 S m^{-1}. They further found that including polylactic acid (PLLA) can result in further increase in electrical conductivity. The excluded volume principle results in selective localization of graphene in PS domains resulting in reduction of the percolation threshold from 0.33 to 0.075 vol%.

Graphene when used as a filler can form an infinite network of connected paths via the insulating matrix. This can result in a rapid increase in electrical conductivity of direct current (DC). The electrical conductivity of such dispersed materials can be described by a bond percolation model (Stankovich et al.[24]). The electrical conductivity of the composite can be given by a power law above the percolation threshold as:

$$\sigma_c = \sigma_f \left(\frac{\varphi - \varphi_c}{1 - l_c} \right)^t \tag{3.2}$$

where σ_c is the thermal conductivity of the composite above the threshold, σ_f is the thermal conductivity of the filler, ϕ is the volume fraction of the filler, ϕ_c is the percolation threshold, and t is the universal critical exponent. The percolation threshold in graphene–PS composites can be seen at about 0.1 vol% graphene. High aspect ratios of graphene sheets can result in a lower percolation threshold. Oblate ellipsoids with an aspect ratio of 1,000 can be predicted to possess a threshold of 0.1 vol%. The antistatic criterion for thin films is met at 0.15 vol%. The graphene sheets have to be incorporated in the matrix in a homogeneous manner. Graphene can be exfoliated in a PS matrix resulting in composites with lower percolation threshold for room temperature electrical conductivity. Graphene sheet dispersions can be achieved with acrylonitrile butadiene and styrene (ABS) matrices. The quality of nanofiller dispersion in the polymer matrix correlates directly with its effectiveness for improving electrical, mechanical, thermal, impermeability, and other properties. The properties of the composite are sensitive to the aspect ratio and surface-to-volume ratio of the filler. Wrinkling and corrugations of the sheets are a consideration. Agglomeration of graphene sheets is not seen even at processing temperatures of 300°C. This is much higher than the T_g, glass transition temperature of PS of about 100°C. Upon preparation, it was determined whether the graphene sheets were single layered or multilayered using TEM.

Ultrahigh-molecular weight polyethylene (UHMWPE) is a macromolecule with molecular weight greater than 1 million. It possesses interesting properties such as good chemical stability, low moisture uptake, low coefficient of friction, high abrasion resistance, and ultrahigh toughness. These are used as implants for use in hip, knee, and spine replacements; bullet-proof armors; wear-resistant belts; and so forth. It is a good insulator of electricity and heat. Nanocomposites can be prepared from UHMWPE and graphite using compression molding. This results in improvement in electrical and heat conduction. Electrostatic spraying of a graphene nanoplatelet (GNP) suspension with an average thickness of 6 to 8 nm and a specific area of 120 to 150 $m^2 gm^{-1}$ has been demonstrated during fabrication of the nanocomposite. Nanocomposite films with greater than 60 graphene sheet stacks have been made with improved fracture toughness and tensile strength at a low GNP content of only 0.1 wt%. In order to obtain better dispersion and high filler content in composites, in situ polymerization was carried out using catalyst supports by Sturzel et al.[25] They used functionalized nanosheets and salicylaldimine catalysts supported on TiO_2, zirconium dioxide (ZrO_2), and CNTs. They found difficulty in achieving ultrahigh-molecular weight during the polymer filling process. Pretreatment of the metallocene catalyst was found necessary. The strategy was to immobilize single-site chromium (III) catalysts on functionalized graphene sheets. It was found that increasing the graphene content in the composite from 5% to 10% by weight resulted in increased stiffness as seen by the increase in Young's modulus of elasticity. Functionalization of graphene was found to be a requirement in order to obtain the increase in stiffness of the product. The reason for this is not clear. Compatibilization of the dispersed and continuous phases needs to be better understood. They found that the functionalized graphene sheets formed more stable dispersions in n-heptane.

Superconductors

Graphene can be used to make organic high-temperature superconductors. Picene is composed of five fused benzenes. Prof. Y. Kubozono at Okayama University in Japan found that the picene crystals when doped with potassium or rubidium atoms[26] exhibit zero resistance at 18 K. The ceramic superconductors exhibit superconductivity at over 100 K temperatures. The 18 K superconducting temperature of doped picenes is similar to the 38 K of potassium-doped Buckminster fullerene and 11 K of calcium-intercalated graphite. Superconductors can be used to make efficient electric motors and power storage and distribution systems. Alkali-doped acenes can make good superconductors. The first materials used to prepare superconductors in the 1980s were mixtures of oxides of copper, yttrium, and barium. It is now believed that Cooper pair electrons are generated at lower temperatures in the material. Once their mutual repulsion is exceeded, the electrons can flow through the material unimpeded. Materials with p orbitals under certain conditions can be made to donate the electrons and effect superconductivity. Molecular superconductors can be made from picenes and acenes. Doped picenes behave in a similar manner as doped fullerenes. An isomer of picene, pentacene, does not superconduct when intercalated with an alkali metal. Perhaps, this is because the pentacene is linear and the fused benzene rings in picene and acene can trigger further motion of electrons. The physics-dominated field of superconductivity is now directed toward chemistry. The use of the cocatalyst methylaluminoxane, along with a catalyst chromium with a functionalized graphene sheet, was found to produce interesting results such as higher catalyst activity, formation of molecular weights higher than 1 million, and superior morphological control. Hydroxy functional groups present at the edges were found to play an important role in the process. Millimeter-size UHMWPE powder particles were produced with a narrow particle size distribution. When functionalized graphene sheets were not used, reactor fouling problems were seen. Bimodal molar mass distribution was seen in the absence of functionalized graphene sheets. The polymer filling technique was found to be effective in making UHMWPE–graphene sheet nanocomposites.

Rapid DNA Sequencing

The HGP was completed ahead of time. By the year 2050, the word "disease" is going to be eliminated as we know it according to scientists at Stanford University, Stanford, CA. Every known function of each organism in the universe is going to be linked to the protein signals. Protein signals are then linked to genes in eukaryotes. M. Schena along with his professor Davis pioneered the microarray analysis method in order to obtain sequencing information from gene expression studies. The genomes of 27 different mammals ranging from giant pandas, African elephants, gorillas, rats, mice, to dolphins have been determined. Genome sequencing will be used to identify drugs that can stop the spread of cancerous cells within the patient. One hundred thousand dollars was the cost for Steve Jobs for his genetic sequencing. Whole-genome sequencing can be ordered these days from outfits such as Knome, Cambridge, MA, for $68,000 and exome sequence for $25,000. Forty-eight thousand dollars is the charge levied by Illumina, San Diego, CA, for whole-genome sequencing. More universities are offering bioinformatics, nanotechnology, and bioengineering as separate branches of study.

The HGP had been completed ahead of time in 2003—in 10 years against targeted 15 years. This involved sequencing 3 billion base pairs. The biological databases double in size every 10 months, and the computing speed of microprocessors doubles in speed every 18 months. So a database search that costs $2 today, 2 years from now would quadruple in cost to $8 on account of the explosive growth of databases and would be cut back in half to $4 on account of the increase in computing power. There is scope for the development of data search and data storage algorithms and methods. It can be viewed as a marriage between information technology and computational biology.[27]

The cost of completion of the first HGP was $2 billion. Advances in DNA sequencing technology have come about that allow genome completion within a month. Reagents needed to sequence a billion base pairs would require as little as $5,000. Other costs such as microarray instrumentation and lab technician fee need be added. As costs sink, the whole-genome sequencing may be ordered for not just research purposes, but also for personal treatment. An example of a person who sought genome sequencing is Stephen R. Quake. A number of his family members responded poorly to anesthetics. A day will arrive in not-so-distant future when every Tom, Dick, and Harry will know details about his personal genomes. Companies are mushrooming that sell whole-genome sequencing services. One of the advantages of whole-genome sequencing is that one does not have to "do it again." When a new disease-causing mutation is discovered, one looks at one's genome and one can tell if one has that mutation.

Research studies are underway to link gene to protein to protein signal to function and hence disease. The exome that forms the protein coding portion of the genome is 1%. Sequence data on the coding portion can be obtained at 20 times lesser cost than that for whole-genome sequencing.

Costs of sequencing are currently higher than costs of isolation of the exome portion of the genome. Hence, cost saving could be achieved in exome sequencing. Jay Shendure of the University of Washington has identified the mutation responsible for Freeman Sheldon syndrome and Miller syndrome. Specialists of gastrointestinal disorders are able to diagnose the disorders rather than making educated guesses before. Shot gun sequencing (SSP) can be used to project the whole genome from the information in the exome. SSP is a not deterministic polynomially bounded (NP) complete problem. It is computationally difficult. Approximate solutions are available. R. K. Wilson at the Genome Sequencing Center at Washington University, St. Louis, is participating in the cancer genome atlas (TCGA). This is a joint effort between the National Human Genome Research Institute and the National Cancer Institute.

They aim to improve the understanding of the molecular basis of cancer through whole-genome sequencing. Current projects are on brain and ovarian cancers. Future projects on breast, kidney, and lung cancers are underway. The personal genome project (PGP) at Harvard University is ambitious in making the sequencing studies more clinically relevant. G. M. Church noted that if the cost of DNA sequencing per base pair continued to follow Moore's law–like progression, scientists would need to start connecting genes and traits. PGP has 15,000 volunteers currently. They have a goal of 100,000 participants. Participants are expected to obtain a perfect score in an entrance examination that demonstrates their knowledge of human genetics and the implications for them and their families of the data being collected. All data from PGP would be made publicly available.

Sanger's method of DNA sequencing was used in HGP. This involves transcription of the DNA template in the presence of dye-labeled modified nucleotides that terminate DNA-strand elongation when they are incorporated. As the modified nucleotides are spread at random in the strands, the sequencing reaction results in a mixture of DNA strands of varied lengths each with its end base labeled with a fluorescent dye. Separation based on lengths of strands can be achieved using capillary electrophoresis. Sanger's method is the gold standard for DNA sequencing. Other less-laborious methods may be used instead of the laborious Sanger's method. Calibration is used in deduction of sequences from the images.

Generalized Fick's law of diffusion–based models can be used to better capture the finite speed of diffusion of the fragments. Mathematical models with improved capability[28] can be used to decrease sequencing errors and improve the efficiency of sequence deduction. Recent laboratory methods developed require less extensive sample preparation and amplification of a library of fragments from genomic DNA. Parallelism is infused in these methods. A genome sequence is put together by alignment of millions of fragments against the reference sequence from HGP. In order to minimize errors, each base pair is identified several times. This is referred to as "fold coverage."

A $1,000 genome is within reach. BioNanomatrix, Philadelphia, PA, works on long strands of DNA. More genetic diversity is available this way. Organization of a genome varies from person to person, all with the same base layout. A nanofabricated device is used to separate double strands with 100,000 to 200,000 base pairs into individual lanes. Blocks of DNA with seven bases are labeled. Location of these blocks forms a bar code for an individual genome. DNA sagamis that appear like smiley faces can be used to design nanorobots and be used in drug delivery and help the surgeon with hard-to-reach locations. Oxford Nanopore Technologies uses a method of reading DNA sequences using nanopores. Bases are identified by the induced charge in the amplitude of the current carried by aqueous ions passing through the pore. An intact DNA strand is threaded through the pore and the bases are identified as they pass through a reading head. This method is slow. Read length depends on the speed or throughput. Further advances in genome sequencing and proteome sequencing can be expected by using SGs. As is, the capability to study more genes per biochip is increasing at a rate much like Moore's law in electronics.

Scientists in the Netherlands claim that they have found a method of rapidly sequencing DNA and RNA. They pass these strands through the nanometer-sized sieves in graphene sheets. A voltage is applied across the sheets. Each of the nucleotide bases—A, G, C, and T for DNA and U in place of T in RNA—has a unique effect on the conductance of graphene as they pass one at an instance of time. Sequence distribution of DNA and RNA is deduced from changes in voltage. Computerization of the procedure and use of sensors with shorter response times can lead to more rapid sequencing of DNA and/or RNA. As each amino acid passes through the hexagonal sieve, the change in electrical characteristics can be used to deduce the amino acid sequence distribution by calibration.

Nanopores can be used to differentiate A, T, G, and C in the DNA sequences. This can lead to a high-speed DNA sequencing method. DNA translocation through graphene nanopores needs to be better understood. Kinetics of DNA passage through the graphene nanopores was studied by Sathe et al.[29] They conducted a series of all-atom molecular dynamics simulations. The effect of voltage variation on DNA translocation was studied by placing a 45 bp ds DNA at the throat of a 2.4-nm graphene nanopore. At the lower bias voltages, the current was blocked more effectively by the DNA. The molecule elongated as the voltage was increased and more ions passed through the pores. Lower bias voltages and negatively charged pores were found to lead to slowing of DNA translocation and partial unzipping of the nucleotide polymer sequence. The polynucleotide was found to stick to the graphene surface. A characteristic double-plateau current signature was found from the simulations of DNA in folded conformation. Individual base pairs can be distinguished in nanopores of ultrathin graphene membranes. Nanopore-based DNA sequencing devices can be developed.

Magnetic Sensors

The need for nanocoatings is pronounced in giant magnetoresistive (GMR) thin film heads, microelectromechanical sensors (MEMS), and fuel cells. GMR thin film heads are used in hard drives of desktop computers. The seven layers are a silicon substrate at the bottom, a tantalum buffer layer, a nickel–iron free layer, a copper spacer, a germanium pinned layer, an iron–manganese pinning layer, and a tantalum gap layer on top. Prototypical magnetoresistive structures include a pair of ferromagnetic layers that are separated by a nonmagnetic spacer layer. The direction of magnetization can change according to the magnetic field. The electrical resistance of the GMR structure depends on the relative orientations of magnetization directions of the free and pinned layers. Data stored in the form of small magnetic fields recorded on a magnetic disk can be detected using the GMR structure. The interleaving insulating layer is a nano-oxide layer.

The problem of magnetic noise is seen in read-head sensors based on GMR and tunneling magnetoresistance (TMR). Extraordinary magnetoresistance (EMR)-based sensors have been proposed by Hitachi Global Storage Technologies, Amsterdam, the Netherlands,[30] for use as read-head sensors in magnetic recording hard disk drives. A graphene sense layer is used. The magnetic field sensitivity and the electrical resistance of the graphene sensor can be tuned by the electric field effect. The highest sensitivity is at the peak of electrical resistance. The charge transport via electrons and holes is caused by the electric field generated by a gate bias voltage that penetrates the graphene sense layer. A super-linear dependence is seen in the response of the sensor as a function of the applied external magnetic field. The state of minimum sensitivity is seen around the magnetic field. A static magnetic biasing field to the graphene sensor is desirable. The need for a graphene magnetic field sensor with a ferromagnetic biasing layer that is used to provide the needed static magnetic biasing field and is located in proximity to the graphene sense layer without short circuiting the graphene sense layer was identified by Hitachi Global Storage Technologies. They have patented a graphene magnetic field sensor with a ferromagnetic biasing layer in proximity to the graphene sense layer. The sensor comprises a suitable substrate, ferromagnetic biasing layer, graphene sense layer, and an electrically insulating underlayer. The underlayer can be made of BN, in a hexagonal lattice. A seed layer is provided to enable the growth of the BN layer. Perpendicular magnetic anisotropy is seen in the ferromagnetic biasing layer. The magnetic moment is in normal direction to the plane of the biasing layer. A graphene sense layer can be made 1 to 10 atom layers in thickness. The graphene field sensor based on EMR may function as magnetoresistive read head in a magnetic recording disk drive. Hexagonal closed-packing (HCP) structures made of alloys of cobalt, nickel, and palladium are examples of materials used in the ferromagnetic biasing layer.

Investments on Nanotechnology in Russia

The Russian initiative plans to spend $8.55 billion in order to create a nanotech industry by the year 2015. In 2004, 85% of global R & D spending on nanotechnology was in the United States, Japan, and the

European Union. London-based consulting firm Cientifica has posted the results of their study in *C & E News* (Vol. 87, 21, 2009, 24). By 2009, the 85% had decreased to 58%. Russia and China are emerging as major players. Russia has taken the second place in nanotech industry size. In 2007, the Russian government started the national nanotechnology initiative. Rusnano is charted with the task of development of a nanotech industry in Russia. This effort is part of a strategy to diversify away from Oil and Gas. This was mentioned in the keynote address at a recent Nanotech conference at Houston, TX, in 2009. Projects shall be financed that lead to nanotech products and auxiliary support services such as certification, standardization, safety, and education. An average of $1.2 billion would be invested every year for a 7-year period from 2007. Loans with a time period of 10 years with "negative interest" rates are offered. In 2008, Russian sales of nanotech products was worth about $545 million. By 2015, the industry size is aimed to grow to about $27 billion. This would be 3% of the world market share. Rusnano provides venture capital, evaluated business plans, and administrative and managerial support. They have a target for revenues and not for profits. Russia is going to collaborate with the United States, South Korea, Israel, and Germany. One year since inception, Rusnano received 1,000 applications for funding. A third was rejected. Half are being reviewed. Twelve proposals have been approved and funded. Eight hundred and sixty-four million dollars is being invested. Some examples of large-scale projects are polycrystalline silicon and monosilane factories in the Irkutsk region. The production of the manufacturing plant is 3,800 tons per year. The products will be used for photovoltaics and semiconductor components. Shift in emphasis from pure to applied research is seen.

Biomedical Applications

There are other biomedical applications besides the rapid DNA sequencing discussed in the section "Rapid DNA Sequencing." Blood compatibility with graphene is an issue in biosensors and drug delivery and cell imaging applications. CNTs have been found to be toxic in biomedical applications. ssDNA can be strongly adsorbed onto graphene sheets. This improves the specificity of complementary DNA. The graphene improves the stability against enzymatic cleavage. Decent electrocatalytic reduction of hydrogen peroxide (H_2O_2) was found in third-generation electrodes made for ssDNA. Biocompatible scaffolds using graphene have been used to delineate mesenchymal stem cells from bone cells still allowing the proliferation. Polyethylene glycol star polymers are covalently grafted onto the edges and activated surfaces of graphene. These structures have been used to impart stability to aqueous, buffer solutions of nanographene oxide (NGO). Even at low contents of graphene, chitosan was found to have improved biocompatibility for L929 cells. Antibacterial graphene paper can be generated. GOs have been found to have dose-dependent toxicity to cells and mice. Cell apoptosis was induced and lung granuloma was formed and the kidney was not able to clean them. The accumulation of graphene in organs such as liver, kidney, spleen, and brain was studied.[31] Graphene samples tested were found to be compatible with blood.

Chemical Modification of Graphene

One of the bottlenecks in the widespread use of graphene in order to realize the promise of two-dimensional single-atomic-layer material is the chemical modification/functionalization of the sheet. This can be achieved by photopolymerization of styrene in the defect sites of graphene sheets. This was shown by Steenackers et al.[32] This is an interesting nondestructive functionalization strategy. These modifications are desirable in sensor and electronic applications. The objective is to provide sufficient chemical activity without changing the electronic transport. Diazonium chemistry may be used to effect chemical grafting on graphene. Polymer brush layers with well-defined functionality, thickness, and density can be formed by surface-induced polymerization on self-assembled monolayers (SAMs). Ultraviolet (UV) photopolymerization of styrene was used in order to grow polymer brushes on graphene sheets. The only requirement for this surface-initiated polymerization is the abstractable hydrogen atoms on the surface. Pristine graphene does not have these abstractable hydrogen atoms. The edge and basal plane defects are tapped to effect the polymerization reactions. This process can be applied to graphene prepared by different methods, such as CVD, on copper, epitaxial graphene on SiC and rGO. Raman spectroscopy was used to confirm that the reactions do not affect the conjugation. Binding sites are limited to the defects. Hall effect measurements are used to confirm that the electronic mobility is not greatly affected because of the photopolymerization. A number of vinyl monomers did not have much reactivity with pristine graphene. Nonreactive monomers can be used to grow polymer brushes using photopolymerization by the method of electron-beam-induced carbon deposition (EBCD). It was shown that polymer brush layers can be covalently attached to graphene by photografting and photopolymerization. This was demonstrated in three different methods of graphene fabrication. Basal plane conjugation of graphene and electronic properties were found to be unaffected after polymerization. AFM was used to confirm the enhanced rates of polymerization with increased number of layers. Self-assembly can be used to build structures into supercapacitors and biosensors. This is a nanoscale covalent chemical modification of graphene.

Nanoscale Thermometry

Graphenes and other nanoparticles can be used to measure a cell's temperature. A laser beam is used to heat gold nanoparticles inside a cell. The differences in fluorescence of nanodiamonds inside the cell can be used to measure the local temperatures. Measurement of nanoscale heat changes can lead to better understanding of biological systems. Gene expression control using temperature and bioheat transfer application needs nanoscale thermometry. Raman spectroscopy and detection of fluorescing proteins can be used for thermal sensing. Most methods used today have poor sensitivity and difficulty with localized measurements. Scientists at Harvard University[33] have developed a nanoscale thermometry using a common defect in diamonds. A nitrogen vacancy center and a nitrogen atom replace adjacent carbon atoms in a diamond lattice. Even very small changes in temperature cause some strain in the lattice of such diamonds. This affects the quantum spin properties of the local defect and the fluorescence properties are modified. These changes are detected leading to a new device. Microwave pulses were used to change the spin states of defects in diamond lattices and the resulting changes in fluorescence are used to obtain corresponding changes in temperature. The resolution of the device can be as good as 1.8 mK in areas of 200 nm across the lattice. The scientists obtained the temperature profile in the cell. The thermometer was tested on an embryonic fibroblast. Nanodiamonds and gold particles were injected into the cell. Changes in fluorescence of diamonds were used to obtain the changes in temperature within the cell. Spatial resolution is comparable to Raman spectroscopy. Real-time biological activities with subcell resolution can be accomplished.

Graphene Nanorobot Drug Delivery Systems

Graphenes and fullerenes can be used in nanorobot drug delivery systems with improved bioavailability.[34,35] About 4,000 mistakes are made by surgeons every year.[36] The brain surgeon cannot reach regions in the cranium such as the hypothalamus on the operating table. Such attempts can increase the morbidity rates for patients who have undergone brain surgery. About 141,000 patients in total with 74,000 male and 67,000 female patients have brain cancer according to the National Cancer Institute's Surveillance Epidemiology and End Results (SEER) program.[37]

Currently, knowledge of the location of the malign cells in the human anatomy is not sufficient to effect cure. Nanorobot drug delivery systems can be a solution to this problem. Advances in the areas of robotics, nanostructuring, medicine, bioinformatics, and computers can lead to the development of nanorobot drug delivery systems. Nanorobots can offer a number of advantages over current methods such as: (1) use of nanorobot drug delivery systems with increased bioavailability; (2) targeted therapy such as only malignant cells being treated; (3) fewer mistakes on account of computer control and automation; (4) reaching remote areas in human anatomy not operatable at the surgeon's table; (5) as drug molecules are carried by nanorobots and released where needed, the advantages of large interfacial area during mass transfer can be realized; (6) noninvasive technique; (7) computer-controlled operation with knobs to fine-tune the amount, frequency, and time of release; (8) better accuracy; and (9) drug inactivity in areas where therapy is not needed minimizing undesired side effects. Fullerenes were mass produced by Frontier Carbon

Corporation (FCC) at 40 tons y^{-1} in 2003. The cost of production of graphenes is expected to decrease in the coming years.

Industrial robots came about in the 1960s. The adoption of robotic equipment came about in the 1980s. In the late 1980s, there was a pull-back in the use of robots. Use of industrial robots has been found to be cheaper than manual labor. More sophisticated robots are emerging. Nanorobot drug delivery systems can be one such robotic system. Seventy-eight percent of the robots installed in the year 2000 were welding and material-handling robots. The Adept 6 manipular had six rotational joints. The most important form of the industrial robot is the mechanical manipulator. The mechanics and control of the mechanical manipulator is described in detail in introductory robotics courses.[38] The programmability of the device is a salient consideration. The math needed to describe the spatial motions and other attributes of the manipulators is provided in the course. The tools needed for design and evaluation of algorithms to realize desired motions or force applications are provided by control theory. Design of sensors and interfaces for industrial robots is also an important task. Robotics is also concerned with the location of objects in three-dimensional space such as its: (1) position and orientation; (2) coordinate system and frames of reference such as tool frames and base frames; and (3) transformation from one coordinate system to another by rotations and translations.

Kinematics is the science of motion that treats motion without regard to the forces that cause it. In particular, attention is paid to velocity, acceleration, and acceleration of the end effector. The geometrical and time-based properties of the motion are studied. Manipulators consist of *rigid links* that are connected by *joints* that allow relative motion of neighboring links. Joints are instrumented with position sensors that allow the relative motion of neighboring links to be measured. In the case of rotary or revolute joints, these developments are called joint offset. The number of independent position variables that would have to be specified in order to locate all parts of the mechanism is the number of *degrees of freedom*.

The end effector is at the end of the chain of links that make up the manipulator. This can be a gripper, a welding torch, an electromagnet, and so forth. Inverse kinematics is the calculation of all possible sets of joint angles that could be used to attain the given position and orientation of the end effector of the manipulator. For industrial robots, the inverse kinematic algorithm equations are nonlinear. Solutions to these equations are not possible in closed form. The analysis of manipulators in motion in the workspace of a given manipulator includes the development of the Jacobian matrix of the manipulator. Mapping from velocities in joint space to velocities in cartesian space is specified by the Jacobian matrix. The nature of mapping changes with configuration. The mapping is not invertible at points called singularities.

Dynamics is the study of actuator torque functions of motion of the manipulator. The state space form of the Newton–Euler equations can be used. Simulation is used to reformulate the dynamic equations such that the acceleration is computed as a function of actuator torque. One way to effect manipulator motion from here to there in a specified smooth fashion is to cause each joint to move as specified by a smooth function of time. To ensure proper coordination, each and every joint starts and stops motion at the same time. The computation of these functions is the problem *of trajectory generation*. A spline is a smooth function that passes through a set or via points. The end effector can be made to travel in a rectilinear manner. This is called Cartesian trajectory generation.

The issues that ought to be considered during the mechanical design of a manipulator are cost, size, speed, load capability, number of joints, geometric arrangement, transmission systems, choice and location of actuators, internal position, and sensors. The more joints a robot arm contains, the more dexterous and capable it will be. But it will also be harder to build and more expensive. Specialized robots are developed to perform specified tasks and universal robots are capable of performing a wide variety of tasks. Three joints allow the hand to reach any position in three-dimensional space. Stepper motors or other actuators can be used to execute a desired trajectory directly. This is called the linear positional control. Kinetic energy of the manipulator links can be calculated using the Lagrangian formulation. Both linear kinetic energy and rotational kinetic energy can be tracked.

Submarine nanorobots are being developed for use in brachytherapy, spinal surgery, cancer therapy, and so forth. Nanoparticles have been developed for use in drug delivery systems, cure of eye disorders, and use in early diagnosis. Research in nanomedicine is underway to develop diagnostics for rapid monitoring; targeted cancer therapies; localized drug delivery and improved cell material interactions; scaffolds for tissue engineering; and gene delivery systems. Novel therapeutic formulations have been developed using poly lactic-co-glycolic acid (PLGA)-based nanoparticles. Nanorobots can be used in targeted therapy and in repair work of DNA. Drexler and Smalley debated whether "molecular assemblers" that are devices capable of positioning atoms and molecules for precisely defined reactions in any environment are possible or not. Feynman's vision of miniaturization is being realized. Smalley sought agreement that precision picking and placing of individual atoms through the use of "Smalley-fingers" is an impossibility. Fullerenes, C_{60}, are the third allotropic form of carbon. Soccer-ball-structured C_{60} with a surface filled with hexagons and pentagons satisfies Euler's law. Fullerenes can be prepared by different methods such as:

1. first- and second-generation combustion synthesis;
2. chemical route by synthesis of corannulene from naphthalene. Rings are fused and the sheet that is formed is rolled into a hemisphere and stitched together; and
3. electric arc method.

Different nanostructuring methods are discussed in the works of Sharma[39,40] These include:

1. sputtering of molecular ions;
2. gas evaporation;
3. process to make ultrafine magnetic powder;
4. triangulation and formation of nanoprisms by light irradiation;
5. nanorod production using a condensed phase synthesis method;
6. lithography;
7. etching;
8. galvanic fabrication;
9. lift-off process for integrated circuit (IC) fabrication;
10. nanotip and nanorod formation by a CMOS process;
11. patterning iridium oxide nanostructures;
12. dip pen lithography;

13. SAM;

14. hot embossing;

15. nanoimprint lithography;

16. electron-beam lithography;

17. dry etching;

18. reactive ion etching;

19. quantum dot

20. sol–gel process;

21. solid-state precipitation;

22. molecular-beam epitaxy;

23. CVD;

24. lithography;

25. nucleation and growth;

26. thin-film formation from surface instabilities;

27. thin-film formation by spin coating;

28. cryogenic milling for preparation of 100- to 300-nm-sized titanium;

29. atomic lithography methods to generate structures less than 50 nm;

30. electrode position method to prepare nanocomposites;

31. plasma compaction methods;

32. direct write lithography; and

33. nanofluids by dispersion.

Thermodynamic miscibility of nanocomposites can be calculated from the free energy of mixing. The four thermodynamically stable forms of carbon are diamond, graphite, C_{60}, Buckminster fullerene, and CNT. Five different methods of preparation of CNTs were discussed. Thermodynamically stable dispersion of nanoparticles into a polymeric liquid is enhanced for systems where the radius of gyration of the linear polymer is greater than the radius of the nanoparticle. Tiny magnetically driven spinning screws were developed. Molecular machines are molecules that can be temporarily lifted out of equilibrium with an appropriate stimulus and can return to equilibrium with regard to the observable macroscopic properties of the system. Molecular shuttles, molecular switches, molecular muscles, molecular rotors, and molecular nanovalves are discussed. Supramolecular materials offer an alternative to top-down miniaturization and bottom-up fabrication. Self-organization principles hold the key. Gene expression studies can be carried out using biochips. Collective nanorobots (CNRs) are a new generation of self-organizing collectives of intelligent nanorobots. This new technology includes procedures for interactions between objects and their environment resulting in solutions of critical problems at the nanoscopic level. Biomimetic materials are designed to mimic a natural biological material. Characterization methods of nanostructures include small-angle X-ray scattering (SAXS), TEM, scanning electron microscopy (SEM), scanning probe microscopy (SPM), Raman microscopy, AFM, and helium ion microscopy (HeIM).

By the year 2050, the number of patients with Alzheimer's disease will be 16 million. The cost of memory loss is about $1.1 trillion. Currently, 5 million patients suffer from Alzheimer's in the United States. The root of *Curcuma longa*, turmeric, has been found to have a positive effect in experiments with rats in curing Alzheimer's disease.[41] The drugs such as bapineuzumab from Eli Lilly and solanezumab from Johnson and Johnson used today are used to target the amyloid-β protein in the brain. This protein has been found to misfold and result in clump formation in patients' brains. Furthermore, these drugs are expensive. Turmeric root is used in curries in India and is readily available from farms at a lower cost. It is yellow colored. This is recommended by Ayurveda and Siddha Medicine. It is known to possess anti-inflammatory and antioxidant properties.

Curcumin (Figure 3.4) has been found to decrease inflammation in the brain. It causes reduction in oxyradicals formed in patients' brains. Frietas[42] defined nanomedicine as follows:

Nanomedicine is the preservation and improvement of human health using molecular tools and molecular knowledge of the human body.

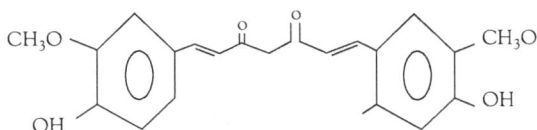

Figure 3.4 Structure of curcumin

Tiny magnetically driven spinning screws were developed by Ishiyama et al.[43] These devices were intended to swim along veins and carry drugs to infected tissues or even to burrow into tumors and kill them with supply of heat. Untethered microrobots containing ferromagnetic particles under forces generated by magnetic resonance imaging (MRI) magnetic fields were tested for travel through the human anatomy at the NanoRobotic Laboratory at Montreal, Canada,[44] in 2003.

A nanorobot to measure surface properties has been patented.[45] Nanorobot technology will prevail 10 to 20 years in the future. The nanorobot unit has a manipulation unit and an end effector. The end effector can be a sensor or made to move as close as possible to the surface of interest. The drive device has piezoelectric drives. The resolution of the measurement can be in the nanometer range and the actual measurement in the centimeter range. The nanorobot is made sensitive in all directions, in multiple dimensions. It can operate under vacuum. Surface roughness can be measured using nanorobots.

A patent was developed[46] for a minimally invasive procedure by means of dynamic physical rendering (DPR). "Intelligent," "autonomous" particles were used. An interventional aid is formed with the help of self-organizing nanorobots. These nanorobots were made of catoms. C-arm angiography is used to monitor DPR procedures. A three-dimensional image data record on the target region is obtained. The determined form was converted to a readable and executable program code for the catoms of the nanorobots. The determined form was transferred to a storage unit. The program code was executed in order to achieve self-organization in the unstructured catoms that were introduced in the target region. The execution of the program was triggered by a timer or position sensor. The intervention aid is used as an endovascular target region.

Nanocrystals with motor properties have been patented.[47] A reciprocating motor is formed by a substrate, an atom reservoir, a nanoparticle ram, and a nanolever. The nanoparticle ram is contacted by the nanolever resulting in movement of atoms between the reservoir and the ram. Substrate and nanolever are made of multiwalled nanotubes (MWNTs) made of iridium. The substrate used was a silicon chip.

The nanoscale oscillator[48] has been patented by Sea Gate technology, Scots Valley, CA. A microwave output is generated by application of a DC that is allowed to pass through layers of magnetic structure separated by nanometer dimensions. Spin momentum transfer (SMT) is a phenomenon realized to exist in 1989. It can be used in magneto-resistive random access memory (MRAM) devices. Phase-locked microwave spin transfer is the next advancement of the technology. The electric current produced is in the GHz spectrum. The local magnetic field source is used in the application of a magnetic field to a free layer of an SMT stack. The magnetic source can be from a horseshoe magnet with poles stationed above and below the SMT stack. The magnetic source can take on other forms such as helical coil that surrounds the SMT or pancake-type coils above and below the SMT stack or an annular pole and a coil that surround the stack. A permanent magnet may be planted above the stack. The SMT stack consists of a top electrode, a free layer, a non-magnetic layer, a pinned magnetic structure, and a bottom electrode.

Nanowhiskers can be grown by control of nucleation conditions.[49] Nanowhisker formation on substrates can be made using the vapor–liquid–solid (VLS) mechanism. A particle of a catalyst material is placed on a substrate and is heated in the presence of gases until it melts. A pillar is allowed to form under the melt. As the melt rises up on top of the pillar, a nanowhisker is formed. Miller direction <111> may be a preferred growth direction of the whiskers. The catalytic property is present at the interface of whisker and air. Indium phosphide (InP) nanowhiskers, for example, were grown using metal–organic vapor phase epitaxy. Characterization of nanowhiskers is done by electron microscopy. MOVPF is a low-pressure metal–organic vapor phase epitaxy process. Fifty nanometers of aerosol gold particles were deposited on an InP substrate. This was placed on a horizontal reactor cell heated by an RF-heated graphite susceptor. Hydrogen was used as the carrier gas. The temperature was ramped to 420°C for 5 min. The molar fraction of the flow rate in the cell was 0.015. Nanowhisker growth was found to commence upon addition of trimethylindium (TMI). The molar fraction of TMI was 3 millionth. The typical growth time for production of nanowhiskers was found to be 8 min.

Design and Control Strategy

The strategy for design of a nanorobot drug delivery system among other things comprises the following salient features:

1. Use fullerenes/graphenes as carriers of curcumin.
2. Complex the fullerene/graphene with curcumin and deactivate the drug action as in other cases of photodynamic therapy.
3. Computer-controlled irradiation can activate the fullerene. Breakage of fullerene and release of curcumin at a constant rate[41] are expected. The intensity, wavelength, and areas of brain that are irradiated are controlled using computer software.
4. Curcumin acts on the diseased cells and cure is effected.
5. A pharmacokinetic model with a fourth-order process for irradiation and competing parallel reactions for fullerene breakage, polymerization of fullerene, and drug action is developed.
6. Time for drug release is computed for the model. The drug delivery model for a fullerene nanorobot curcumin delivery system is created.
7. Chromophores in curcumin are used as a dye and picked up from imaging. Information from the sensor on drug action is compared against model and feedback control, effected by changing the intensity and wavelength of irradiation. This is used to trigger feedback control or model-based control of irradiation using computers.
8. Intrathecal route of entry for the nanorobot fullerene curcumin complex can be preferred to sublingual entry. The choice of the cerebrospinal fluid system may obviate any digestive disorders from fullerene carbon.

The following can be concluded from the study;[50,51]

1. Robotics has emerged as a collegiate course about 20 years ago. A leading supplier of robots had made over 100,000 robots by the year 2001. Advances in the field of robots over a 30-year period include the weight of the robot made having increased to 500 kg. About 40 different nanostructuring operations were reviewed recently. The vision of nanorobots is a natural next step. Recent developments in nanomedicine were reviewed.

2. Principles from photodynamic therapy, fullerene chemistry, nano-structuring, X-rays, computers, pharmacokinetics, and robotics are applied in order to develop a design methodology for nanorobot drug delivery of *Curcuma longa* for treatment of Alzheimer's disease.

3. Nanorobots can offer a number of advantages in drug delivery over present methods. These include more bioavailability, targeted therapy, fewer surgeon mistakes, reaching remote areas in human anatomy, large interfacial area for mass transfer, noninvasive technique, computer control of delivery, better accuracy, less side effects, and greater speed of drug action.

4. The design includes the use of fullerenes as carriers of *Curcuma longa*. The drug is complexed with fullerene and deactivated. Intrathecal injection of fullerene and drug is preferred to sublingual entry in order to minimize digestive side effects. Chromophores in curcumin are picked up from imaging. Feedback or model-based control of irradiation using computers is effected. Computer-controlled irradiation activates the drug by breakage of fullerenes. The intensity, wavelength of radiation, and areas of brain that are activated are controlled by a computer. Time for drug release is computed from the pharmacokinetic model. The process is assumed as a fourth-order process for irradiation and competing parallel and series reactions for fullerene breakage, polymerization of fullerene, and drug action.

5. Drug action can be viewed as a set of nanoreactors operated as plug flow reactors (PFRs). The transient dynamics of a PFR is studied in detail. The governing equation for the transient dynamics in a PFR is given in the work by Sharma (2013a)[52] and can be seen to be a hyperbolic partial differential equation (PDE). Wave concentration can be identified in the model equation. The steady state concentration is given in Sharma (2013a).[53] Wave characteristics were found in the solution of the model equations. It can be seen that there is a wave component to the transient concentration that can be delineated and studied as shown by Sharma (2013a).[54] The model solution for transient concentration of the drug is given in Sharma (2013a)[55] and plotted for typical values of Damkohler number.

6. At the time of the reaction between drug and target, other reactions may come about such as polymerization of curcumin, polymerization

of fullerene, and dissociation of the curcumin under the irradiation conditions. The yield of drug action has to be optimized. In order to optimize the yield, the Denbigh scheme of reactions is applied to the system under consideration. The model equations that can be used to describe the dynamics of the five reactant/product species in a continuous stirred tank reactor (CSTR) can be written in the state space form and given in Sharma (2013a).[56] The eigenvalues are negative and hence the system is stable.

7. The model equations are solved by method of Laplace transforms. The model solutions for the five reactant species in the Denbigh scheme is given in Sharma (2013a).[57]

8. The conditions where instability may arise and the types of instability are provided in Sharma (2013a).[58]

9. A reaction scheme with three reactions in series and three reactions in parallel is shown in Sharma (2013a).[59] This may be applicable when the dissociation of *Curcuma longa* is also considered. A sparse K matrix was seen. This system was found to be an integrating system.

Summary

The market for nanotechnology is expected to reach $3 trillion by the year 2015. The market for graphene is expected to rise to $675 million by the year 2010. Scalable production methods for graphene are being developed. Graphene can be used to effect further speed increases higher than 30 pHz of microprocessor speeds. Prof. P. Kim at Columbia University and coworkers clocked an electron and hole mobility of 120,000 cm^2 V^{-1} s^{-1} in annealed SG at 240°K. The intrinsic mobility of graphene in 200,000 cm^2 V^{-1} s^{-1}. This can lead to the development of ultrafast electronics. Narrower gate widths are achievable using graphene. IBM clocked 26-GHz speeds in graphene transistors in 2008. A gate length of 240 nm was achieved at Watson Research Center, NY. Use of graphene has been shown in further miniaturization efforts. MLGs are developed at University of California (UC) Riverside that can result in accelerated database searching, data compression, and image recognition. Computer architecture that obviates the von Neumann bottleneck is expected. Tunable bandgaps using bilayer graphenes were shown in UC Berkeley and can be used in nanophotonic and nanoelectronic devices. Stable carbon chains with 16 carbon atoms can be used as the basic component in electronic devices.

Scientists at the University of Texas at Austin have paved the way for ultracapacitors by showing that electrical charge can be stored in an SG structure. Graphene sheets can be used to rapidly store and discharge EC to develop rechargeable batteries and fuel cells. Aggregation of sheets can be a problem in manufacture of electrodes. Transformation of graphene sheets to crumbled-ball structure has resulted in improved scalability of graphene-based ultracapacitors.

Molecular dynamics simulations using the LAMMPS package were used to study the desalination dynamics of graphene materials. The water flux rates and salt rejection rates can be expected to be higher for desalination membranes made using graphene compared to those used currently in sea water reverse osmosis (SWRO) sea water RO plants. Desalination performance was found to be sensitive to pore size and pore chemistry. CDI technology can be used to generate potable water from the sea. Lower energy consumption, ready regeneration, and better environmental friendliness can be expected. Electrosorption can result in the formation of strong double layers of ions on the electrodes. Electrosorptive performance can be increased by use of GNFs.

Samsung has patented LEDs using p-type and n-type doped graphenes. Boeing has patented an MMC, with graphene nanoplatelets with increased thermal conductivity ($713 \text{ W m}^{-1} \text{ K}^{-1}$) for use in heat spreaders and other thermal management devices. Transparent graphene films can be used to make flexible and stretchable electrodes. Use of graphene layers in solar cells can lead to higher light to electricity conversion as high as 15%. Graphene can be used in rechargeable batteries that can be used in electric vehicles. Graphene has been used to make the thinnest coating for protecting metals against corrosion. Same amount of corrosion protection can be obtained using SG as that was obtained using five layers of organic coatings. At Neurocare in Europe, carbon-based implants in ears, eyes, and the brain are evaluated. In the navy, graphenes can be used as infrared transparent, electromagnetic shielding coating for electro-optic sensor windows and domes. High electrical conductivity and optical transparency were the attributes of graphene tapped into.

BN nanosheets called "white graphene" can be used to abate water pollution. Thirty-three times greater absorption potential is used to clean up oil spills. Surface modification of graphene can result in fabrication of electronic devices with better electrical conductivity and strength. Graphene can be used in polymer nanocomposites as an intercalant along with UHMWPE. Graphene has been used to make organic high-temperature superconductors. Doped picene and potassium-doped fullerenes have been shown to have higher superconducting properties. Graphene sheets can be used to obtain sequence distribution in DNA and RNA rapidly. Scientists at the Netherlands pass the DNA strands through nanopores in graphene and apply a voltage across the sheet. Sequence distribution information is deduced from the changes in voltage. EMR-based sensors have been proposed by Hitachi Global Storage Technologies, Amsterdam, the Netherlands. Biomedical applications of graphene have been shown in biosensors and drug delivery and cell imaging applications. Photopolymerization of styrene was used to grow polymer brushes on graphene. Chemical modification is an important step for widespread use of graphene.

Rusnano is an $8.55 billion initiative to create a nanotech industry by 2015. A nanoscale thermometer can be developed using the changes in fluorescence characteristics of defects in a diamond lattice. Nanorobots can be developed that effect cure of disorders that are difficult to treat. Principles from photodynamic therapy, fullerene chemistry, nanostructuring, X-rays, computers, pharmacokinetics, and robotics are applied in the design of nanorobots for treatment of Alzheimer's disease. The *Curcuma longa* that has shown curative effects in rats' brains with Alzheimer's is complexed with fullerenes. The drug is inactive when caged. It is infused intrathecally into the cerebrospinal system. Irradiation of the hypothalamus and other areas of the brain where Alzheimer's disease is prevalent leads to breakage of fullerenes and availability of the drug with the diseased cells. Due to better mass transfer, better cure is effected.

CHAPTER 4

Stability

Chapter Objectives

- Interfacial stability
- Euler stability
- Island growth
- Epitaxial substrate—atomic bonding
- Scroll stability
- Free energy of reaction
- Thermodynamic stability
- Edge stability/hole formation
- Defects
- Buckling and fracture
- Metastability

Continuous Monolayers and Epitaxial Substrates

Many attempts have been made in order to prepare ultrathin films. For all practical purposes, the collective experience of several investigators indicates that continuous monolayers are nearly impossible to make. When a metal film that is few nanometers in thickness is evaporated, the film is found to become discontinuous. Islands/holes are formed in the material structure. Island growth has been attributed to minimization of surface energy. Epitaxial substrates can be used and their contribution in Eqn. (4.1) is against the surface energy contribution.

Two-dimensional crystals cannot be grown in isolation without the presence of an epitaxial substrate. The epitaxial substrate provides the additional atomic bonding needed. It is known that the density of thermal fluctuations of a two-dimensional crystal in multidimensional space diverges with temperature. This is also called the Landau–Peierls argument. About 75 years ago, Landau and Peierls provided proof that the two-dimensional crystal as a matter of general rule is not thermodynamically stable. Thermodynamic fluctuations of such crystals lead to atomic displacements that are comparable in size to the interatomic distances at any finite temperature. Mermin in the late 1960s extended this argument and the theory was refined as the Landau–Peierls–Mermin theory. Geim[1] estimated the limit in possible sizes L of two-dimensional atomic crystals. The divergence with temperature of density of thermal fluctuations is logarithmic. Crystal growth requires higher temperature at which the atoms are more mobile and the lattice is softer with less shear rigidity.

$$L \sim ae^{\frac{E}{T_G}} \tag{4.1}$$

where a, lattice spacing, is ~1 A°; E, atomic bond energy, is 1 eV; T_G is the growth temperature. Estimates of graphene at room temperature lead to astronomical sizes. Self-assembly may allow for growth of graphene at room temperatures. The thermal fluctuations are quenched by use of an epitaxial substrate. The Landau–Peierls argument is obviated and the island formation is prevented. When graphene is grown on graphite, the interaction is weak.

Peierls transition states that perfect order in a one-dimensional crystal is broken when a distortion of the periodic lattice occurs and atomic positions vibrate. One-dimensional equally spaced chains with one electron per entity are inherently unstable. Proof has been provided in the form of a simple model. The potential for an electron in a one-dimensional crystal with lattice spacing a is considered in the model. The Kronig–Penney model can be used to explain bandgaps in semiconductors. Energy bandgaps are created in the E–k diagram at multiples of $k = \pi/a$ on account of the periodicity of the crystal. Experimental evidence of such dimerization or transition was presented at Stanford University in 1964 by W. Little. He theorized that a certain class of polymer chains may experience a high T_c superconducting transition. Lattice distortions can lead to pairing of electrons as per the Bardeen–Cooper–Schrieffer (BCS) theory of superconductivity to be replaced. Electrons may form Cooper pairs instead of forming ions. The transition temperature can be improved by a factor given by the square root of the ratio of the mass of the ion to that of the charged particle responsible for distortions. This factor can be as high as 300. Peierls transition can result in an insulating transition in some materials. Ion cores in Peierls transition on account of fluctuations in electron density can generate charge density waves. Examples of materials where charge density waves have been found are $NbSe_3$ and $K_{0.3}MoO_3$ at temperatures of 145 K and 180 K. Peierls transition can result in transformation of polyacetylene to graphene. First-principles calculations were used to confirm the Peierls transition of graphene nanoribbons (GNRs) from polyacetylene. The calculations were performed using the Gaussian 03 package with periodic boundary conditions[2] using density functional theory (DFT). Spin polarized effects were not considered.

Euler Stability

The intrinsic flatness of graphite was reevaluated by Kroto.[3] In the scale of a few atoms of carbon in a hexagonal sheet form, edge instabilities form in the way of dangling bonds. These dangling bonds lead to a spheroidal cage structure in the case of fullerenes. Euler's law may be applied to structures where energy-driven closure can be achieved with the inclusion of non-hexagonal disclinations. Twelve pentagonal disclinations can lead to closure resulting in graphene balloons. In highly ordered pyrolytic graphite (HOPG) materials, the presence of H and OH groups in the edges of the polycyclic aromatic precursor facilitates the formation of graphene sheets and interplanar bonding. During the formation of extended structures, nanoscale restructuring of edges takes place that plays a salient role in the formation of the structure of the carbon allotrope.

Thermodynamic Stability—Free Energy of Reaction

Free energy of formation estimates of carbon chains indicate a thermodynamically unfavorable state for formation of graphene sheets that have less than 6,000 carbon atoms. About 24,000 carbon atoms are required for the two-dimensional geometry to be favorable over the various three-dimensional configurations. The ceiling temperature phenomenon in polymers was recently discussed.[4]

The thermochemistry of graphene formation can be better understood using first-principles thermodynamics-based nucleation size estimation. Why growth on copper substrates is better than others can be explained. Control of nucleation size is an important method to improve graphene quality and productivity. Zhang et al.[5] combined electronic structure calculations, molecular dynamic (MD) simulation, and thermodynamic analysis to better understand the observed growth process of graphene on copper surfaces. They found that carbon atoms on a copper surface are unfavorable under typical experimental conditions. The active species for graphene growth are largely CH_3, CH_2, and CH groups. Nucleation behavior can be used to explain many experimental observations. During synthesis of graphene using chemical vapor deposition (CVD), it has been found that the carbon atoms dissolve in a nickel substrate. The growth process is limited to the surface of copper. Copper is thus a superior choice as a substrate. Sample quality is sensitive to experimental conditions used. A two-step process with low- and high-pressure stages has been reported in order to produce large single-sheet graphenes.

The atomic process or the mechanism of reactions when graphene grows on a copper substrate is not well understood. Hydrocarbon decomposition and graphene nucleation on a copper surface were studied from first-principles thermodynamics.[6] Electronic structure calculations were performed using DMOL software where DFT is implemented by use of PBE exchange-correlational functional. The numerical basis used was double-ζ numerical basis set with polarization function (DNP) and DFT semicore pseudopotential for all atoms. Space cutoff of 4.4 A° and thermal smearing of 0.002 hartree were used. Copper with Miller surfaces of (111) and (100) was used in the simulations of the dehydrogenation reactions with a five-layer p(3×3) slab with a ~15 A° vacuum. Geometry optimization was performed with all atoms relaxed except those in the bottom two layers that were kept at their bulk positions. A $4 \times 4 \times 1$ k-point grid was adopted. All calculations were performed in a spin-polarized frame. C_{24} and C_{54} graphene clusters were used in the carbon chemical potential during adsorption on copper surfaces. A p(6×6) supercell was used. Transition state search was performed with synchronous transit methods. First-principles MD simulations were performed using the Vienna ab initio simulation package (VASP) code for CH-covered (111) copper surface and (100) copper surface with a three-layer p(6×6) surface model. The

CH coverage was set to 0.33 ML on the (111) surface and 0.42 ML on the (100) surface. Only k-point sampling was used. The canonical ensemble MD simulation was performed at 1,300 K for 4.0 ps and 1.0 fs. The temperature reached equilibrium after 0.5 ps. The production trajectory lasts for 3.5 ps. The dehydrogenation kinetics studied comprised the following elementary steps:

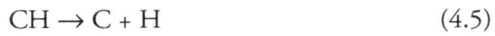

$$CH_4 \rightarrow CH_3 + H \tag{4.2}$$

$$CH_3 \rightarrow CH_2 + H \tag{4.3}$$

$$CH_2 \rightarrow CH + H \tag{4.4}$$

$$CH \rightarrow C + H \tag{4.5}$$

The initial step is the adsorption of methane on a copper surface. The final step is the adsorption of carbon atoms and four hydrogen atoms at

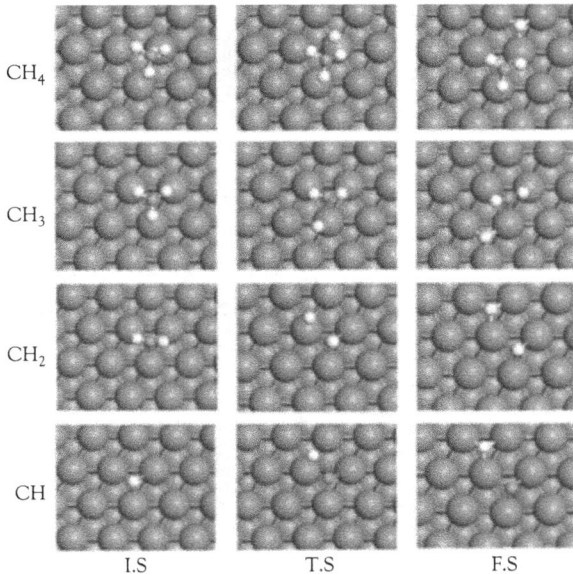

Figure 4.1 Geometric structures of the initial, transition, and final states of the four steps of dehydrogenation

the surface. Geometric structures of initial, transition, and final stages of the dehydrogenation reactions are shown in Figure 4.1. Reaction energies, activation energies, bond lengths, and lattice parameters were monitored for all the four CH_x species. Reactions given in Eqns. (4.2) to (4.5) are endothermic. The activation energy barriers are about 1 to 2 eV. The final products of atomic carbon and atomic hydrogen are 3.6 eV higher in energy compared with adsorbed methane. This means that atomic carbon is energetically unfavorable on a copper surface. The energy of atomic carbon is more stable on the (100) copper surface compared with the (111) copper surface. The copper coordination number of atomic carbon is higher on the (100) surface. Similar endothermic behavior is found for ethylene on the (111) copper surface.

Decomposition of methane can be exothermic on active metal surfaces such as palladium or ruthenium. In addition to the reactions shown in Eqns. (4.2) to (4.5), other reactions among the CH_x species may take place. MD simulations were performed on a CH-covered (111) copper surface at 1,300 K. Significant surface structure relaxations were found during the 3.5-ps trajectory. No CH dissociation was observed. A strong thermodynamic force to recombine carbon and hydrogen atoms can be seen. Ethylene decomposition on a (111) copper surface is a more favorable reaction path compared with CH dissociation. CH recombination on a (111) copper surface is exothermic. C–Cu–C bridge structures are formed. On account of these structures, higher energy barriers for ethylene + CH reaction are formed. Copper lattice distortion on the (100) surface is lower compared with that on the (111) surface. Cu–C interactions are stronger than Cu–Cu interactions. Diffusion of CH groups is slower on the (100) surface. Complete dehydrogenation of CH_4 is not favorable.

The chemical potential of hydrogen was calculated at experimental growth temperature of 1,300 K and reference pressure of 1 bar. One half of the energy of hydrogen molecules was used as reference according to National Institute of Standards and Time—Joint Army Navy Air Force (NIST-JANAF) thermochemical tables.

$$\mu_H\left(T, P_0\right) = 0.5\left(h\left(T, P_0\right) - h\left(0, P_0\right) - TS\left(T, P_0\right)\right) \qquad (4.6)$$

Under arbitrary pressure,

$$\mu_H(T,P) = \mu_H(T,P_0) + \frac{k_B T}{2}\ln\left(\frac{P_{H_2}}{P_0}\right) = -0.975 + 0.056\ln\left(\frac{P_{H_2}}{P_0}\right) \quad (4.7)$$

The free energy change of reaction for methane is:

$$\Delta G_{CH_4}(T,P_0) = h(T,P_0) - h(0,P_0) - TS(T,P_0) = -2.87 \quad (4.8)$$

Reference values of half the energy values of hydrogen molecules and carbon atoms were taken and the DFT calculated energy of methane was -9.23 eV. Thus:

$$G_{CH_4}(T,P) = -12.1 + 0.112\ln\left(\frac{P_{CH_4}}{P_0}\right) \quad (4.9)$$

When methane and molecular hydrogen are pumped in simultaneously, the ratio of the partial pressures of methane and hydrogen can be taken as χ. The relationship between chemical potentials of atomic carbon and atomic hydrogen at the equilibrium of methane and molecular hydrogen can then be written as follows:

$$\mu_C = G_{CH_4} - 4\mu_H = -2\mu_H - 10.15 + 0.112\ln(\chi) \quad (4.10)$$

The left hand side (LHS) of Eqn. (4.10) can be calculated from the experimental partial pressure of hydrogen and the ideal gas approximation.

At a fixed ratio of partial pressures of methane and hydrogen, χ, chemical potential of atomic carbon, μ_C, during CVD growth on a copper surface can be related to the partial pressure of molecular hydrogen. This is shown in Figure 4.2. It can be seen that when a surface carbon species has chemical potential higher than μ_C, it is unstable and will react with molecular hydrogen. From Figure 4.2, it can be seen that the chemical potential of an isolated atomic carbon on a (111) copper surface is much higher than μ_C and hence is unstable. This is so for the (100) surface and typical experimental conditions. This expression can be corrected by addition of an adsorption energy such as 0.01 eV/c for both (111) and (100) copper surfaces. This provides the thermodynamic driving force

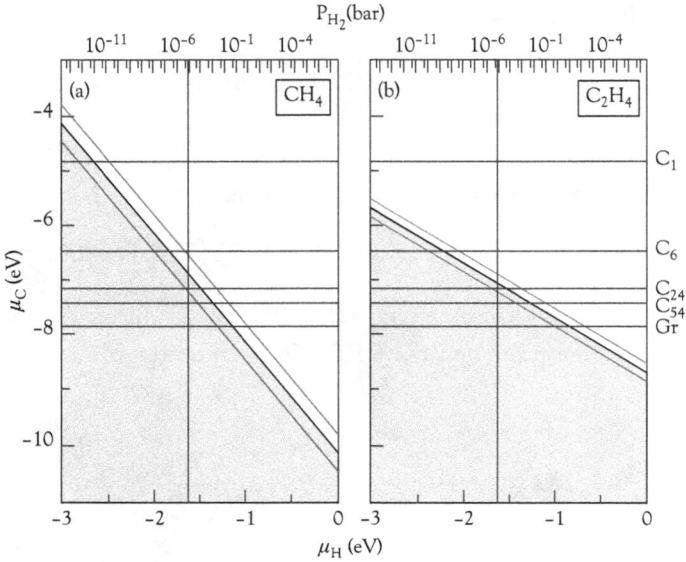

Figure 4.2 *Relationship between chemical potential of atomic carbon and partial pressure of hydrogen during growth of graphene on a (111) copper surface at 1,300 K with methane*

for growth of graphene. At high pressures as well the carbon can become unstable. Infinite two-dimensional graphene is stable on the surface, whereas the smallest carbon cluster is unstable.

When does the transition from unstable to stable graphene happen? The nucleation size of graphene growth can be calculated. The potential energy of a carbon cluster of n atoms adsorbed can be written as follows:

$$\mu_C^n = \frac{1}{n}\left(E_{C_n/\text{surf}} - E_{\text{surf}}\right) - E_C \tag{4.11}$$

The terms of significance in Eqn (4.11) are energy of the adsorbed system ($E_{Cn/\text{surf}}$), energy of the copper surface (E_{surf}), and the energy of an isolated carbon atom in vacuum (E_C). Calculations[7] for one atom for clusters of C_6, $C_{24,}$ and C_{54} on (111) and (100) copper surfaces were completed. From the calculated potential energies, the nucleation size can be determined. The nucleation size was estimated as six carbon atoms for partial pressures of hydrogen and methane at 10^{-5} bar and 2×10^{-4} bar, respectively, for the (111) surface. For $\chi = 0.05$, the nucleation size can be increased to 24. For ethylene, Eqn. (4.10) can be calculated as:

$$\mu_C = -\mu_H - 8.7 + 0.056 \ln(\chi) \qquad (4.12)$$

Identification of active surface species is a salient consideration in graphene growth. Small carbon clusters are more stable compared with single carbon atoms. Low solubility of carbon in copper is a reason for surface-limited growth as seen in experiments. Nucleation size is an important parameter to control graphene quality. Growth rate increases when hydrogen concentration decreases. This may be explained by smaller nucleation sizes at higher χ values. Lower χ leads to lower nucleation sizes and better quality of graphene. Partial pressure of methane has therefore to be kept on the lower side. Atomic carbon was found to be unstable on the copper surface under different experimental conditions.

Scroll Stability

Larger-size graphene sheets are susceptible to scroll up into carbon nanotubes (CNTs). Estimates of the size when this happens can be obtained by accounting for competing contributions from the bending and surface energies. When placed on a substrate, van der Waals interactions may be sufficient to prevent graphene from scrolling.

Booth et al.[8] prepared graphene membranes of macro-sizes of ~100 μm diameter. They characterized the samples using transmission electron microscopy (TEM). The graphene sheets were supported on one side. These samples were found not to scroll. They were able to withstand loads millions of times their weight. The substrates were made out of SiO_2 or 90-nm-thick polymethylmethacrylate (PMMA). The graphene film is protected using a copper film brought in place using electrodeposition. They found that when the substrate was fragmented by annealing, the graphene sliver left extended 10 μm in the absence of an external support. This observation was counterintuitive to the belief that graphene would scroll in order to minimize the excess energy in the free surface and dangling bonds. Suspended graphene has been reported to have scrolled edges. The graphene sheet is seen to possess high stiffness. It is found to scroll when suspended in liquid but not in air. In the study by Booth et al. (2008),[9] the elasticity theory has been applied and the effective thickness of single-layer graphene, "a" can be calculated as:

$$a = \sqrt{\frac{\kappa}{E}} = 230 \text{ pm} \tag{4.13}$$

Two hundred and thirty picometers is smaller than the carbon–carbon bond length of 1,420 pm. In Eqn. (4.14), the bending rigidity κ at room temperature and Young's modulus of elasticity E used were 1.1 eV and 22 eV A^{-2}. Estimation of Young's modulus of elasticity is made from the bulk modulus of elasticity of graphene. The length-to-thickness ratio of a graphene sheet is about a million. When the graphene sheet extends from the support, it means that the rigidity of the material is rather high. The density of graphene is ~ 7.6 E -7 kg m^{-2} and surface area is $\sqrt{3}d^2/4$.

Booth et al.[10] considered a sheet of graphene that was infinitely thin with length l and width w anchored about the y-axis and free bending under gravitational acceleration, g m.s^{-2}. The total energy of the sheet, E, is given by:

$$E = \frac{\kappa}{2}w\int_0^l \left(\frac{d^2h}{dx^2}\right)^2 dx - \rho gw\int_0^l h dx \tag{4.14}$$

where x is the distance from the anchor point at $x = 0$ and $h(x)$ is the deviation from the horizontal axis that remains the same along y. The deviation $h(x)$ found by minimization of energy and application of boundary conditions is:

$$h(x) = \frac{\gamma l^2 x^2}{4} - \frac{\gamma l x^3}{6} + \frac{\gamma x^4}{24} \tag{4.15}$$

where $\gamma = \rho g/\kappa$ ~5 E13 m^{-3} and ρg ~ 7.5E-6 N m^{-2}. The maximum bending angle can be seen at $(dh/dx) = \gamma l^3/6$ at ~l = 20 µm. Equation (4.15) was arrived at neglecting in-plane stresses. In the case of graphene, bending is limited due to high in-plane stiffness. Apparent rigidity is determined by stretching rather than by simple bending. Out-of-plane deformation was estimated as:

$$\frac{h}{l} = \sqrt[3]{\frac{\rho g l}{E}} = \sqrt[3]{3E - 14l} \qquad (4.16)$$

l and w in the estimates and square sheets were considered. From the estimates, gravity-induced bending is found to be ~10^{-4} for the graphene slivers observed in Booth et al.'s study[11] and the corresponding value of plain strain is 10^{-8}. Copper nanoparticles can be expected to have 100 times more strain. For a collapse of the graphene membrane discussed in the study by Booth et al.,[12] an acceleration of 10^6 g m.s^{-2} is required. Hence, graphene membranes are intrinsically stiff. They are often corrugated. The corrugation has an effect of increase in rigidity of graphene.

Interfacial Stability

Solidification is considered as one of the most important manufacturing processes. The casting process is used to produce several million pounds of steel, copper, zinc, and aluminum alloys. At some point during their processing, most metalized materials, ceramics, glasses, and thermoplastics are in their liquid state. As the temperature is lowered below the freezing point, the molten materials tend to solidify. During the solidification of materials that crystallize the atomic arrangement, they are found to change from short-range order (SRO) to long-range order (LRO). Two steps can be associated with the transformation. The first step comprises formation of nuclei of solid phase from the liquid. These nuclei are ultra-fine crystallites. The second step is the growth of the nuclei as atoms from the liquid attached to the nuclei. This can proceed till the liquid runs out. Only some crystallization process ends up in LRO. Others end up as partial crystalline materials or materials that are amorphous.

Some of the principles of solidification that have been found to be true for other materials may still hold good for two-dimensional graphene materials. The term nucleation refers to the formation of the nanocrystallites from the molten material. This can include solution precursors as well. In general, nucleation refers to the initial stage of formation of one phase from another phase.[13] This can be written as follows:

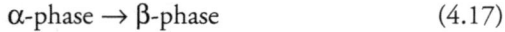

$$\alpha\text{-phase} \rightarrow \beta\text{-phase} \tag{4.17}$$

The β-phase can be the two-dimensional graphene sheet and α-phase can be the vapor phase or precursor material depending on the process type as discussed in the section "Thermal Management" in Chapter 3. When the solid forms, a solid–vapor interface is created. A surface free energy σ_{sv} is associated with this interface. Let the thickness of the two-dimensional sheet of graphene that is formed be given by "t" and the surface area of the sheet on one side be given by "A." The volume of the crystalline material would then be $V = At$. The total change in free energy can be written as follows:

$$\Delta G = At\,\Delta G^v + A\,\sigma_{sv} \tag{4.18}$$

where ΔG^v is the free energy change per unit volume. This has to be negative since the phase transformation is thermodynamically feasible. The surface free energy can be assumed to be not a strong function of thickness "t" and hence to be a constant during the phase transformation process. It has the units of energy per unit area. ΔG^v does not vary with t. An embryo is a tiny sheet of solid that forms from the vapor as carbon atoms cluster together. The embryo is not stable and may either grow into a stable nucleus or vaporize.

Equation (4.18) is shown for two-dimensional graphene in Figure 4.3. It can be seen that there exists a certain sheet thickness "t" below which the interface is unstable from free energy considerations. This may be the reason for use of copper foils and transfer of foils in some of the synthetic methods discussed in Chapter 5. The volume free-energy change starts at 0 and decreases linearly as the thickness is increased. The surface free energy is relatively constant and is independent of the sheet thickness. The total free-energy change of the phase transformation as given by Eqn. (4.14) for a certain value of ΔG^v and σ_{sv} is shown by the curve in Figure 4.3. A stable system is formed when the total free energy change of the transformation is negative. The numerical values of the critical thickness depend on the solid–vapor interfacial tension of graphene and free-energy change

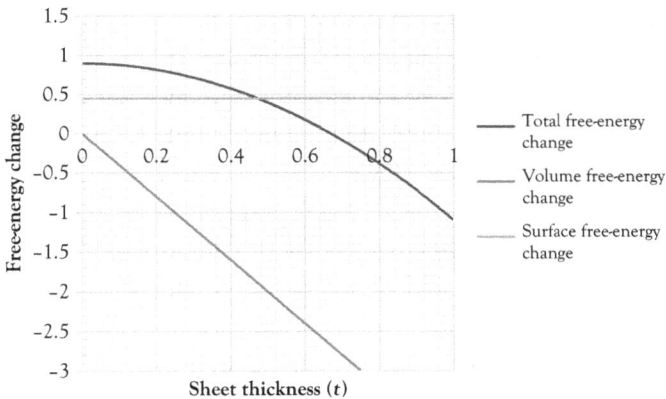

Figure 4.3 Interfacial stability—critical thickness of a sheet. Figures obtained at a surface tension of 0.9, $\Delta G(v) = -8$, and $A = 0.5$

per unit volume of the precursor. These values are difficult to obtain from experimentation. This can be predicted from equilibrium considerations.

Edge Stability

The behavior of atoms at a boundary such as the edges is a concern. In the section "Interfacial Stability," the atoms on the interface were discussed and in the section "Continuous Monolayers and Epitaxial Substrate," the behavior of atoms from two dimension to three dimension was considered. The mechanism of edge reconstruction and the stability of "zigzag" edge configuration were studied by Girit et al.[14] The study of graphene edges is expected to have large impact on electronic properties of GNRs. It has been difficult to resolve the boundaries of graphene sheets at the atomic level, and characterization of one-dimensional interface of a two-dimensional sheet is a challenge. Typical problem areas in using a scanning probe microscope (SPM) and an atomic force microscope (AFM) for edge characterization include (1) scanning speed of several minutes prevents study of dynamics of atoms in a few seconds; (2) better resolution found at cryogenic temperatures when movement of atoms are frozen out; and (3) presence of a substrate can change the dynamics of the atoms studied. Girit et al.[15] achieved sub-angstrom resolution using a transmission electron aberration corrected microscope (TEAM) and mapped every atom in the two-dimensional lattice. The dynamics of carbon atoms in the edge of a hole was studied using a TEAM at 80 kV. The preparation method for a microscopic specimen was discussed in an earlier study. The experiment was conducted under vacuum (10^{-7} mbar) in a microscopic chamber. Movies were filmed that were used to capture the evolution of the hole. Each frame averaged 1 s of average exposure and were 4 s apart. Spatial sampling was 26 ± 4 pm/pixel. The mechanism of hole formation was captured in four figures. One of them is shown in Figure 4.4.

The hole was initially formed by prolonged irradiation by the electron beam near the center of the frame and was surrounded by a hexagonal carbon lattice. Structures limiting the boundary of the frame are adsorbates that were deposited during the graphene sheet formation. The mechanism by which the hole changes shape is given in the three figures in the study by Girit et al.[15] The metastable configuration of the edge is recorded by a film. Ejection of atoms from the edge is induced by the beam and is contrasted from that of the surface as sputtering. The energy barrier for migration of carbon atoms to various neighboring vacant sites is expected to be less than 15 eV necessary sputtering.

The evolution of graphene holes was simulated using the kinetic Monte Carlo method. Edge dynamics was described using three mechanisms:

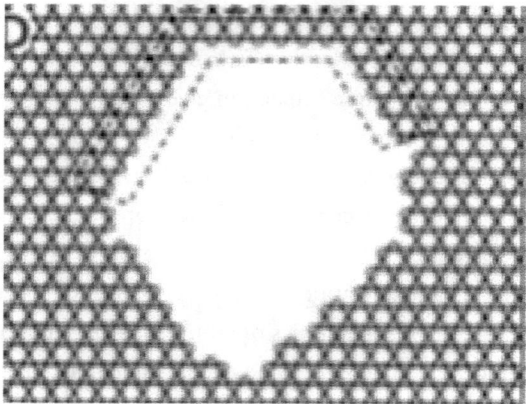

Figure 4.4 Edge configuration by an aberration-corrected TEM image of 19 hexagon zigzag edges interrupted by two 60° turns

(1) beam-induced ejection of carbon atoms along the edge; (2) addition of carbon atoms from a virtual reservoir; and (3) migration of dangling carbon atoms from one site to another. The probabilities of the occurrence of the three mechanisms were calculated. The probability of addition of carbon atoms was found to be smaller than the probability of ejection of carbon atoms. Probability of migration is the least of the three. Holes were found to have order such as "armchair" or "zigzag," in the edges. LRO was found from the simulations. It was determined from the simulations that long armchair edges are less prevalent than zigzag edges. The shape of the hole was found to change with the beam-induced ejection of atoms and resultant migration and edge reconfiguration. The growth of the hole as a function of time was studied. It was found that the zigzag configuration was more stable. Transformation from zigzag to armchair configuration is difficult where the edge is aligned with the zigzag direction of the lattice. A simple model can be used in order to account for the stability of zigzag edges observed in both experiment and simulation by consideration of the effect of ejecting an atom at the edge for each chirality. Fifty percent of the atoms along a zigzag edge are bonded to two neighboring atoms and the rest are bonded to three neighboring atoms. The stabilities of the zigzag and armchair configurations were quantitated. Reconfiguration occurs on a time scale on the order of seconds. The long-term stability of zigzag edges is explained by a simple model and time-average analysis.

Metastability

Another important thermodynamic stability consideration is the metastability. Gibbs free-energy change of formation of graphene structures has to be negative for stable structures. Unstable structures may be expected when the second derivative of the Gibbs free energy with respect to phase volume is negative. Metastable structures can result when the Gibbs free-energy change is negative and the phase stability criterion is not met. Graphene sheets with less than 24,000 atoms/25 nm are metastable. Thus:

$$\Delta G \leq 0 \text{ (favorable for formation)} \qquad (4.19)$$

$$\partial^2 \Delta G > 0 \text{ (phase stability)} \qquad (4.20)$$

Metastability arises when the first condition (Eqn. (4.19)) is met and the second condition is not met (Eqn. (4.20)).

Defects

The imperfections in the arrangement of atoms and ions in three-dimensional space affect the material properties. Some of this is applicable to the two-dimensional graphene sheet material. Three types of defects can be seen:

1. Point defects;
2. Dislocations along the line; and
3. Surface defects

Dislocations have been found to strengthen metals and alloys but are undesirable in crystalline silicon used in computer chip manufacture. Defects are sometimes intentionally introduced. During the process of carburization of steel, small amounts of carbon are introduced in the lattice structure of iron. Transformed into an alloy, carbon steel can be seen to possess higher strength. Doped silicon with n and p junctions can be used to make diodes and transistors. Different layers of n and p junctions can lead to thyristors, inverters, barristers, and so forth. Consider a crystal of pure alumina. This is colorless and transparent. Addition of a small amount of chromium in the form of defects in the crystal results in a ruby crystal. Transistors and the advancement of packing transistors in silicon chips have led to the computer revolution and the information technology revolution. Impurities in copper can lead to reduction of the resistivity of copper and are not desirable in microelectronic applications. One type of defect is the grain boundary. Under certain conditions, mixtures of oxides can act as ceramic superconductors, that is, electrical conduction close to infinite conductance. The presence of grain boundaries increases the electrical resistance. Grain boundaries have been found to be useful in increasing the strength in metallic materials.

Localized interruptions in an otherwise perfect atomic or ionic arrangement are called point defects. Each defect can involve one or more atoms/ions. It can be a missing atom, an extra atom out of sequence, a larger substitutional atom, a smaller substitutional atom, or a missing ion. A missing atom is called a vacancy. An extra atom out of sequence is the interstitial atom. Frenkel defects are seen in interstices and a missing ion with charge is a Schottky defect. Defects can be caused by heating and occur during processing of the materials and introduction of impurities. When impurities are introduced intentionally, they are called dopants. Boron and phosphorus dopants in known concentrations are introduced in the silicon material to tailor the electronic properties of the material.

When vacancies are created, the entropy of the material increases. Vacancies are created during solidification, at high temperatures, high pressures, or when subjected to radiation dosages. The concentration of vacancies increases in an Arrhenius manner with temperature:

$$\frac{N}{N_v} = e^{-\frac{E_v}{RT}} \tag{4.21}$$

The left hand side of Eqn. (4.21) is the ratio of the number of vacancies to the number of atoms per unit volume, E_v is the energy required to make one mole of vacancies in cal mol^{-1}, R is the molar gas constant in cal mol^{-1} K^{-1}, T is the temperature in Kelvin. One vacancy per 1,000 atoms may be found near the melting temperature. The equilibrium concentration of vacancies at a given temperature and pressure is given by Eqn. (4.21).

An interstitial defect is formed in a crystal structure when an extra atom/ion is found among otherwise periodic perfect arrangement. The surrounding region of the defect is distorted. Carbon atoms are added intentionally. Impurities sometimes creep in. The concentration of interstitial atoms remains the same independent of the temperature.

Substitutional defects are those when different types of atoms/ions replace the otherwise periodic, perfect arrangement in the lattice. The size of the substitutional entity can be either larger or smaller than the repeating atom/ion. When the size is larger, the surrounding space is cramped and when the size is small, there is void space formed in the surrounding. The substitution can be either by accident as impurity or deliberate as dopant/alloyant. Once introduced, the number of impurities remains the same and is independent of the temperature. Doping can be effected by chemical or electrochemical methods. Examples of substitutional impurity are arsenic, boron, phosphorous in silicon, or occasionally gallium. Ceramic materials are not defect free. When magnesium oxide (MgO) is added to cupric oxide (CuO), some Mg^{++} ions occupy Cu++ sites. The size of the ions/atoms determines whether the guest is substitutional or interstitial. Valency of the species is also a salient consideration. Point defects can be introduced in graphene in order to increase its magnetic response. This would enable use of graphene in charge and spin manipulation. Point defects in graphene have been shown to carry magnetic moments and they can couple ferromagnetically. Nair et al.[16] showed that point defects in the form of fluorine (F) atoms in a graphene sheet carry magnetic moments with spin ½. Defects lead to paramagnetism. No ordering of magnetic domains was found at temperature ranges down to liquid helium temperature. Induced paramagnetism is the principal contributor to the magnetic properties of graphene at low temperatures. The magnetic response achievable was ~1 moment, μ per 1,000 atoms. Clustering of dopants leads to structural instability of graphene and hence a plateau on the magnetic response. Pristine graphene is diamagnetic. Spin-1 paramagnetism has been reported for ion-implanted graphitic nanoflakes. Kondo effect and negative magnetoresistance have been seen in doped graphene devices. Tunneling density peaks when vacancies are formed in graphite.

A Frenkel defect is a vacancy–interstitial pair when an ion jumps from a lattice point to an interstitial site. Frenkel defects can be seen in graphene sheets. The Schottky defect occurs when a stoichiometric amount of ions is missing from the sheet.

Line imperfections are called dislocations. Deformation and strengthening mechanisms in composites can be explained using dislocations. These may be formed during the process of solidification. Screw dislocations are those when the sheet is skewed by one-atom spacing. Screw dislocation can be identified by construction of Burgers vector. Dislocations are line defects that begin to move when force is applied to the material and the deformation process of the material begins. The critical resolved shear stress is the stress required to move the dislocation. The dislocation moves in a slip system. The system comprises slip plane and slip direction. Slip direction is usually in the close-packed direction. Slip plane is close-packed or near close-packed. Number of slip directions and slip planes are a critical factor in the performance of graphene materials. The ductile behavior can be explained. The ductile-to-brittle transition phenomenon and the temperature associated with it are also investigated. Point defects, both vacancies and interstitial atoms, introduce compressive or tensile strain fields and perturb the atomic arrangement in the crystal. Dislocations therefore cannot slip in the vicinity of point defects, and the strength of the material is enhanced. Surface defects happen at grain boundaries. Smaller grain sizes increase the grain boundary area. Dislocations cannot readily pass through the grain boundary. The strengthening of the material can be given by the Hall–Petch equation. The types and number of crystal defects determine the ease of movement of dislocations. These play a direct role in mechanical performance of the material. Defects have a significant effect on the electrical, optical, and magnetic properties of materials.

Edge dislocation in graphene (section "Edge Stability") sheets can be an extra row of carbon atoms in otherwise repeating, perfect, and hexagonal structure. Both screw and edge can combine and form the mixed dislocation. Two types of stresses are of interest during analysis of the effect of dislocations on material properties. These are normal stress and shear stress. The normal stress (pressure) is the force applied in the direction perpendicular to the plane divided by the area of the plane. Shear stress is the tangential force to the plane divided by the area of the plane. When sufficient shear stress is applied, a process called slip can take place in a graphene sheet. The bonds across the slip plane are broken. A slip plane may be defined as the dislocation line and the Burgers vector as in the three-dimensional case. Atoms below the slip plane are shifted and ready to bond with the atoms in the edge dislocation. Continuation of the process results in the movement of the dislocation through the two-dimensional crystal. Breakage and reformation of bonds require less energy compared with instantaneous breakage of all bonds across the slip plane. The stress required to cause the dislocation to move from one equilibrium position to another is called the Peierls–Nabarro stress and is given by:

$$\tau = ce^{-\frac{kd}{b}} \tag{4.22}$$

where d is the interplanar spacing between adjacent slip planes and b is the magnitude of Burgers vector. c and k are constants characteristic of the material.

Dislocations can be used to explain a mechanism of plastic deformation. Irreversible deformation or change in shape of the material when the stress that caused it is removed is called plastic deformation. The applied stress causes dislocation motion that results in irreversible deformation. Plastic deformation is different from elastic deformation where the change in shape is temporary and reversible. Elastic deformation may be due to stretching of interatomic bonds, and no dislocation motion takes place. The slip process is a salient consideration in understanding the mechanical performance of graphene and materials in general. Slip can be used to explain ductility. Drawing graphene into wires is an example. Mechanical properties of the material can be controlled by introduction of obstacles to the motion of dislocations. Dislocation density in graphene may be reported as length of dislocations per unit area. Dislocations may affect electronic and optical properties of the material. Resistance of copper increases with increase in dislocation density. However, dislocations have a deleterious effect on the performance of photodetectors, light-emitting diodes, lasers, and solar cells. Compound metal oxide semiconductor (CMOS) devices such as GaAs and AlAs can be seen to have dislocations. Here the origin of dislocations is the inequalities in concentration in the melt from where the crystals are grown. It may be due to thermal gradients and induced stresses that the crystals are exposed to during cooling from the growth temperature. In graphene, the slip line may be identified in a two-dimensional sheet material.

The properties of graphene as discussed in Chapter 6 are sensitive to defects. Ripples in graphene sheets can induce pseudo-magnetic gauge fields. Some of the defects reported in graphene[17] are as follows:

1. Pentagon–heptagon pairs
2. 5–7, 7–5 Stone–Wales (SW) defects
3. Pentagon–octagon–pentagon Divacancies
4. 9–5, nonagon–pentagon asymmetric vacancy
5. 5–7–7–5 and 7–5–5–7 adjacent pairs
6. Symmetric vacancy in boron nitride, λ, μ

Bonilla and Carpio[18] considered a planar graphene sample and used Navier–Stokes equations of linear elasticity for the two-dimensional displacement vector (u,v). They ignored vertical deflections and considered in-plane deformations in the continuum limit. They introduced a phenomenological damping coefficient, γ:

$$\rho_2\left(\frac{\partial^2 u}{\partial t^2}\right) + \gamma\left(\frac{\partial u}{\partial t}\right) = \mu\left(\frac{\partial^2 u}{\partial y^2}\right) + (\lambda + 2\mu)\left(\frac{\partial^2 u}{\partial x^2}\right) + (\lambda + \mu)\left(\frac{\partial^2 v}{\partial x \partial y}\right)$$

(4.23)

$$\rho_2\left(\frac{\partial^2 v}{\partial t^2}\right) + \gamma\left(\frac{\partial v}{\partial t}\right) = \mu\left(\frac{\partial^2 v}{\partial x^2}\right) + (\lambda + 2\mu)\left(\frac{\partial^2 v}{\partial y^2}\right) + (\lambda + \mu)\left(\frac{\partial^2 u}{\partial x \partial y}\right)$$

(4.24)

where ρ_2 is the two-dimensional mass density and μ and λ are called the Lame coefficients in two-dimension. Equations (4.23) and (4.24) can be seen to be hyperbolic partial differential equations. Sharma[19] has presented a number of analytical solutions without violation of the second law of thermodynamics for hyperbolic partial differential equations applied to damped wave heat conduction. Bonilla and Carpio[20] showed a procedure to construct defects by numerical simulation. They used stable cores corresponding to the far field of a single edge dislocation and a single dislocation dipole. They compared the numerical solutions with experimental results with graphene and other two-dimensional materials. More complex defects can result when two dislocation dipoles are constructed. Equations (4.23) and (4.24) were discretized on the hexagonal lattice and rewritten in primitive coordinates and finite difference equations replaced with periodic functions. a, b, and c are the hexagonal lattice parameters. The characteristic time is the time taken for a longitudinal sound wave to traverse a distance a. Damping time was also estimated in terms of parameters, λ, μ, γ, and a.

The characteristic time was about 10 fs and the damping time was about 10 s from the simulation of a SW defect to disappear after creation by irradiation. Far from defect cores, for small differences in displacement vector, the resulting discrete equations are reduced to those of continuum linear elasticity. Damping terms are added and the governing equations were solved for appropriate initial and boundary conditions that are consistent with known solutions of edge dislocations. Numerical solutions of these equations are used in order to explain the stability and evolution of experimentally observed defects in suspended graphene sheets. Observed defects are cores of edge dislocations, dislocation dipoles, or pairs of dipoles. Pentagon–heptagon pairs are isolated dislocations. Dislocation dipoles may be symmetric vacancies, nonagon–pentagon pairs that are asymmetric vacancies, 5–8–5 divacancies, 5–7–7–5 SW defects, and 7–5–5–7 defects. Pairs of dislocation dipoles are 5–7–7–5 SW defect adjacent to a 7–5–5–7 defect. The metastable defect is one with three pentagons, three heptagons, and one hexagon. Correspondence between numerical simulations and experimentally observed defects is high.

Buckling and Fracture

Examples of mechanical instability include buckling under tension and fracture on compression for graphene sheets. Thermal strain can induce buckling and rippling in graphene sheets. A sign of the strain such as compression or tension from a theoretical standpoint is not important. Practical applications are used to obtain the compression and tension limits that the graphene sheet can endure. When the film fractures on application of tensile stress, two fracture surfaces are created. The critical tensional strain for fracture can be calculated by equating the strain energy to the energy of the fractured surfaces:

$$\varepsilon_{cr}^{t} \sim 2\sqrt{\frac{\gamma}{El}} \qquad (4.25)$$

where E is the Young's modulus of elasticity, γ is the surface energy, and L along with the uniaxial tension is applied. Equation (4.25) is independent of film thickness. Compressive stresses may induce two types of instabilities, that is, fracture and buckling. For thicker films, the critical compressive strain for fracture is comparable to critical tensional strain for fracture. For thinner films, buckling is seen to precede fracture. Critical compressive strain is determined by the competition between bending energy and stretching energy. Bending energy is given by 0.5 BAκ² and stretching energy is given by 0.5 EtεAm². Bending modulus B is ~Et^3 and t is the thickness of the film. κ and m are curvature and slope due to out-of-plane deformation. The critical compressive strain is $\varepsilon^c_{cr} \sim t^2$. It can be seen that the critical compressive strain is strongly dependent on the film thickness. Thinner films are expected to buckle more. Symmetry of response to compressive and tensile strain changes as the material is made thinner. A graphene sheet is one-atom-layer thick. Greatest asymmetry in strain response between compression and tension can be expected for graphene materials. Continuum mechanical calculations and MD simulation were used by Zhang and Liu[21] in order to investigate different forms of mechanical deformation of graphene induced by uniaxial and biaxial compressive and tensile strain. Narrow GNRs were considered under uniaxial strain along its lateral direction. Buckling under compression and fracture under tension were seen. The total strain energy can be written from the continuum mechanics theory by adding the contributions from bending and stretching mechanisms as follows:

$$U_{tot} = \frac{B}{2} \int_A \left(\frac{\partial^2 \zeta}{\partial x^2} \right)^2 dA + \frac{Et}{2} \int_A \varepsilon^c \left(\frac{\partial \zeta}{\partial x} \right)^2 dA \qquad (4.26)$$

where $\zeta(x)$ is the out-of-plane displacement at any point along the ribbon x. Energy minimization using variational method and application of periodic boundary condition leads to the following expression of critical compressive strain for buckling instability:

$$\varepsilon^c_{cr} = \frac{\pi^2 t^2}{3 \left(1 - v^2 \right) L^2} \qquad (4.27)$$

where v is the Poisson ratio and L and t are the length and thickness of the GNR, respectively. The critical tensile strain for fracture can be written as:

$$\varepsilon^t_{cr} \sim 2 \sqrt{\frac{E_{edge}}{ELt}} \qquad (4.28)$$

where E_{edge} is the edge energy of GNR. Equations (4.27) and (4.28) were also derived from MD simulations. For representative values of Young's modulus of 4.27 TPa, Poisson ratio of 0.19, thickness of sheet of 700 pm, and E_{edge} of 1 eV A^{-1} and 1.2 eV A^{-1} for armchair edge and zigzag edge, the critical compressive and critical tensile strains were plotted in by Zhang and Liu (2011)[22] as a function of the length of the sample. For lengths of 10 to 1000 nm, the critical compressive strain varied inversely with the square of the length of the sample and the critical tensile strain varied inversely with the square root of the length of the sample. The critical compressive strain for buckling was found to be four orders of magnitude smaller than the critical tensile strain for fracture. Maximal asymmetry can be seen between compression and tension modes. Orthogonal buckling for thicker samples is also discussed. Graphene sheets are expected to be elastically isotropic. Similar results as in the GNRs were found in the trends of critical compressive strain. When biaxial compressive strain is applied, the longer side of the graphene sheet can be expected to buckle first and the shorter side will not buckle as long as the sheet is rectangular. Buckling may be simultaneous in length and width directions for a square sheet. The analytical solution for the critical biaxial compression for buckling is:

$$\varepsilon_{x,bi} = \frac{\pi^2 t^2}{3\left(1 - v^2\right)L'^2}$$

(4.29)

where $L'^2 = (L_x^2 + L_y^2)/2$ when the x and y direction lengths are nearly equal to each other.

Summary

Continuous monolayers without discontinuity are difficult to make. Islands/holes are formed. An epitaxial substrate is needed to grow two-dimensional crystals and provide the additional bonding that is needed. Landau and Peierls proved that two-dimensional crystals are thermodynamically unstable. Thermal fluctuations comparable in size to interatomic distances lead to atomic displacements. Equation (4.1) is the Geim's estimate of the limit of size of two-dimensional crystals to L. Thermal fluctuations are quenched by the use of an epitaxial substrate. The Kronig–Penny model can be used to explain bandgaps in semiconductors. A certain class of polymer chains may experience high critical temperature (T_c) superconducting transition. Polyacetylene can be transformed to graphene using Peierls transition. Euler stability considerations were shown by Kroto in his work on fullerenes that few pentagons are needed in order to close a sheet of hexagons. Dangling bonds may lead to the formation of spheroidal cage structure such as fullerenes. Twelve pentagonal disclinations are needed for Euler closure and formation of graphene balloons.

Formation of graphene sheets with less than 6,000 carbon atoms is thermodynamically not favorable. Approximately 24,000 carbon atoms are required for the two-dimensional geometry to be favored over a three-dimensional configuration. Thermochemistry of graphene formation can be used to explain observations such as a copper substrate being more suited for the growth process of graphene. Carbon atoms dissolve in a Nickel substrate. Control of nucleation size was found to improve quality of graphene and productivity. DFT was implemented using DMOL software on the computer. Copper with Miller surfaces of (111) and (100) was used in the simulations of the dehydrogenation reaction with a five-layer slab spaced 15 A° apart. Elementary steps in the dehydrogenation kinetics studied are given in Eqns. (4.2) to (4.5). The important steps in the process are adsorption of methane on a copper surface, exothermic decomposition of methane, Cu–C interactions, surface diffusion, and CVD growth of graphene on a copper surface. The chemical potential and free energy expressions for the intermediate steps during the formation of graphene are written. The transition point of unstable to stable graphene can be estimated from the potential energy of a carbon cluster of n atoms. Active surface species is a salient consideration.

Large-size graphene sheets are preferred from thermodynamic stability considerations. However, larger size graphene sheets scroll up into CNTs. By accounting for competing contributions from the bending and surface energies, estimates of CNTs can be made. Van der Waals interactions with the substrate are sufficient for prevention of scrolling. Booth et al. performed some estimates[23] of the total energy of the graphene sheet (Eqn. (4.14)). By energy minimization, maximum bending angle was estimated. High in-plane stiffness was found to cause bending in graphene. One million g is needed for collapse of graphene. Corrugations can be found and the rigidity is increased.

Formation of two-dimensional graphene from vapor phase can be seen as a phase change reaction (Eqn. (4.17)). The surface energy of the solid–vapor interface and the total change in free energy for the formation of a graphene sheet can be calculated using Eqn. (4.18). Free-energy change per unit volume needs to be negative for the graphene sheet formation to be stable. Eqn. (4.18) is shown in Figure 4.3. It can be seen that there exists a certain sheet thickness "t" below which the interface is unstable due to surface free-energy considerations (Eqns. (4.19) and (4.20)).

Girit et. al.[24] studied the stability of zigzag edge configuration. Edges can have a large impact on electronic properties. Characterization of a one-dimensional interface of a two-dimensional sheet is a challenge. A TEAM was used to map every atom in the two-dimensional lattice. The mechanism of hole formation is captured in Figure 4.4. Shape change of holes and growth of hole as a function of time were also studied. A simple model for long-term stability of zigzag edges was developed.

Metastable structures can result when the Gibbs free-energy change is negative and the phase stability second derivative of free energy consideration is not met.

Point defects, surface defects, and dislocations along the line are applicable for two-dimensional graphene sheets. Localized interruptions in an otherwise perfect atomic or ionic arrangement are called point defects. Sometimes, defects can be used profitably such as in carburization of steel and doping of silicon. Although it can strengthen alloys, dislocations are not desirable in a crystalline structure. Vacancy is caused by a missing atom. Frenkel defects are seen in interstices, and missing of ions is called a Schottky defect. Equation (4.21) is the Arrhenius relationship for increase of concentration of vacancies with temperature. Interstitial defects and substitutional defects are discussed. Implications in doping and magnetization of materials are discussed. Dislocations can be used in order to explain deformation and strengthening mechanism. Sheets can be skewed resulting in a screw dislocation. An extra row of carbon atoms can lead to edge dislocation. Mixed dislocations are a combination of screw and edge dislocations. Slip can arise from sufficient shear stresses. Movement of dislocations from equilibrium position can be given by Peierls–Nabaro stress (Eqn. (4.22)). Plastic deformation, ductility, electronic, mechanical, and optical properties can be explained using dislocations. Slip can damage the performance of LEDs and CMOS devices. Six different defects have been found in graphene by Girit et al.[25] pentagon–heptagon pairs, Stone–Wales defects, pentagon–octagon–pentagon divacancies; nonagon–pentagon vacancy, 5–5–7–5 and 7–5–5–7 adjacent pairs; and symmetric vacancy.

Navier–Stokes equations of linear elasticity with phenomenological damping coefficient were applied to planar graphene by Bonilla and Carpio.[26] Sharma[27] has presented analytical solutions for hyperbolic partial differential equations that can be used to obtain solution of Eqns. (4.23) and (4.24). Damping time was estimated by Bonilla and Carpio (2011)[28] and defects were simulated on the computer.

Buckling and fracture are examples of mechanical instability. Critical tensional strain for fracture is given by Eqn. (4.25). Continuum mechanical calculations and MD simulations were used by Sharma[29] to better understand mechanical deformation. Total strain energy is given by addition of contributions of bending and stretching (Eqn. (4.26)). Expressions for critical compressive strain for buckling, critical tensile strain for fracture, and critical biaxial compression for buckling are provided in Equations (4.27) to (4.29).

CHAPTER 5

Fabrication Methods

Chapter Objectives

- Cost of production
- Roll-to-roll transfer process after synthesis in APFR
- Solvothermal reduction using NMP
- Exfoliation from a carbonizing catalyst
- Ion implantation and layer thickness control
- Reduction of ethanol
- Phase separation of carbon–metal melts
- Unzipping of CNTs
- Electrophoretic deposition and reduction
- Coal pitch as carbonizing source
- Gas intercalation and exfoliation
- Nanoribbon alternation
- Flash cooling
- Shell formation
- Fullerenes
- Carbon nanotubes
- Electrochemical method
- Other substrates: Fe, Ru, Co, Rh, Ir, Ni, and Au

Cost of Production

Currently, the cost of production of graphenes is on the higher side. The cost depends on the substrate used. Copper substrates are less expensive compared with other substrates. A 50 × 50 mm, monolayer thin film of graphene from Graphene Square costs $250 for copper substrates and $808 for PET substrates. Graphene nanoplatelets (5- to 8-nm thick) are sold at $218 to $240 per kg by XG Sciences. A graphene sheet (GS) may have an area with length greater than or equal to 1 mm and fall in a range of 1 to 1,000 mm along the transverse and longitudinal directions. The world's largest graphene factory has begun operations in August 2013. Ningbo Morsh Technology has established a graphene line with a per-year capacity of 300 tons. In the next 5 years, China plans to build graphene infrastructure in Chongqing municipality with revenues of 100 billion yuan ($16.4 billion).

Unscrolled carbon nanotubes (CNTs) form graphene. Two-dimensional nanosheets cannot be generated without an epitaxial substrate. The substrate is used to provide atomic bonding in the third dimension (Landau–Peierls argument). This is an important stability consideration when attempting to synthesize a material with the thickness of an atom. Raman spectra are used to confirm the monolayer structure of graphene.

Another important thermodynamic stability consideration is the metastability. Gibbs free-energy change of formation of graphene structures has to be negative for stable structures. Unstable structures may be expected when the second derivative of the Gibbs free energy with respect to phase volume is negative. Metastable structures can result when the Gibbs free-energy change is negative and the phase stability criterion is not met. GS with less than 24,000 atoms/25 nm are metastable.

The honeycomb lattice structure has been confirmed for graphenes. A transmission electron micrograph (TEM) of Graphene X^{TM} is shown in Figure 1.1. The hexagonal arrangement of atoms is striking and hard to miss! The hexagonal sheets of atoms are planar. The electrons in the sp^2-hybridized orbitals get delocalized. The delocalization of electrons in a monolayer GS is shown in Figure 1.2. Electrons from the top row of hexagonal rings flow to the second layer due to delocalization. This phenomenon was first observed in benzene and was called the Kekulé structure. The discovery was made in the year 1865. Alternating single and double bonds can be seen in the hexagonal rings. In order to maintain the octet configuration, some hexagonal rings have three alternate single and three double bonds and some hexagonal rings have two double bonds and the rest single bonds in an alternating manner. All carbons have an octet configuration. Further reduction of unsaturation may be attained by polymerization by scientists. Electrons can flow readily without any obstacle. This can lead to interesting electrical properties of the material.

Different morphologies of graphene are possible in addition to the sheet structure such as chiral, armchair, and puckered.

Synthesis in an Annular Plug Flow Reactor and the Roll-to-Roll Transfer Process for Larger Areas

Thick/thin films of graphene can be made on flexible substrates. Rigid substrates are not needed. The chemical vapor deposition (CVD) method is used to synthesize graphene and then it can be rolled into thin films by a transfer process. Rectangular graphene films with dimensions of 30 inches across the diagonal were prepared in South Korea in 2010. The process[1] comprises three steps:

1. Adhesion of polymer supports to the graphene on a copper foil: Two rollers are used to get the graphene film grown on a copper foil to be attached to a polymer film coated with an adhesive film as it passes through.
2. Etching of copper layers: Electrochemical reaction with an aqueous 0.1 M ammonium persulphate solution, $(NH_4)_2S_2O_8$, enables the removal of copper layers.
3. Release of the graphene layer onto a target substrate: Thermal treatment is used to detach the graphene from the polymer support and reattach the film onto a target substrate. This target substrate could have been placed below the copper foil in order to obviate the third step.

Three Steps

Graphene is synthesized in a annular plug flow reactor (APFR). The annular reactor space is generated by an 8-inch outer quartz tube and an inner 7.5-inch copper foil-wrapped quartz tube. The use of the annular reactor in place of the tubular reactor is to minimize radial temperature gradients. This was found to cause inhomogeneity in the film formation. The inner tube is heated to 1,000°C. Hydrogen is allowed to flow at 8 sccm and 90 mTorr. The annealing process comes next. Annealment for 30 min allows for increase in grain size in the copper foil from a few microns to 100 microns. This has been found to increase the graphene growth. Methane (CH_4) is allowed to mix with the flowing hydrogen at a flow rate of 24 sccm at 460 mTorr. The sample is rapidly cooled at about $10°C\ s^{-1}$. The graphene film grown on the copper foil is attached to a thermal release tape by applying pressure on the rollers at 0.2 MPa. The copper foil is etched in a plastic bath filled with an etchant. The etched film is washed with deionized water to remove the unused etchant. The graphene film is ready for transfer to a target substrate such as a curved surface. One hundred and fifty to two hundred millimeter per minute transfer rates by thermal treatment can be achieved by letting the graphene film pass through the rollers at a mild heat of 90 to 120°C. Multilayered graphene can be made by repeating this process. The product that comes as a result of this would be different from the bilayer or multilayer material formed during the reaction by other methods. This can be viewed as physical stacking of the formed graphene layers. Screen printing can be used to generate four-wire touch panels.

Continuous production of graphene on a large scale is possible. Scalability of the process is high. Processability is good. Carbon has limited solubility in copper even at 1,000°C. The copper may have a catalytic effect on the graphene formation reaction. Transparent electrodes can be made using graphene on a large scale and replace the currently used indium tin oxide (ITO) electrodes. Monolayer graphene structure was confirmed using Raman spectra. Bilayer and multilayer islands were found using atomic force microscope (AFM) and TEM. Stacked layers reduce the optical transmittance by 2.2 to 2.3% a layer and the conductivity also decreases. Dopants can be added as desired. A p-junction formation by doping can be achieved by addition of nitric acid (HNO_3). Sheet resistance can be increased by chemical doping. Polymethylmethacrylate (PMMA) can be used as a polymer support. One of the challenges in using this method is the formation of polycrystalline graphene due to the occurrence of nucleation again to form a second layer. Oxidation of copper has to be avoided. High rates of evaporation of copper from the foil can hinder graphene growth. Copper is not as effective as nickel to lower the energy barrier to form graphene.

Other carbon sources in addition to methane that can be used for the purpose of preparing graphene are carbon monoxide (CO), ethane (C_2H_6), ethylene (C_2H_4), ethanol (C_6H_5OH), acetylene (C_2H_2), propane (C_3H_8), butane (C_4H_{10}), butadiene (C_4H_6), pentane (C_5H_{12}), pentene (C_5H_{10}), cyclopentadiene (C_5H_6), hexane (C_6H_{14}), cyclohexane (C_6H_{12}), benzene (C_6H_6), toluene (C_7H_8), iso-butane, iso-pentane, and hexene. Mixtures of these sources may also be used. Carbon formation from carbon sources is thermodynamically favored only at high temperatures. The temperature that can be used in the reactor is 300 to 2,000°C. The flow rate range can be 5 to 1,000 sccm. Inert gases can prevent undesirable oxidation. Reynolds number in the reactor is in the regime where laminar flow can be assumed. The reactor is operated at low pressure. This would enable the entire vapor and solid system to fall near the sublimation curve of the of the pressure–temperature (P–T) diagram of reactants and products.

APFR Performance

A horizontal low-pressure chemical vapor deposition reactor (LPCVD) has been discussed by Fogler (2006).[2] The reactor is operated at 100 Pa. The APFR discussed in the previous paragraph (Figure 5.1) is operated at about 60 Pa. One advantage of using LPCVD is the capability of large number of wafers without sacrifice to film uniformity. At low pressures, the diffusion coefficient is expected to increase. Sometimes, Knudsen diffusion effects cannot be ignored. The surface reactions are expected to be rate limiting compared with other mass transfer effects. Assume that the reaction mechanism for the formation of graphene on a copper foil in the APFR is as follows

Dissociation

$$2CH_4 \rightarrow C_2H_2 + 3H_2 \tag{5.1}$$

1,000°C
Adsorption

$$C_2H_2 + 2Cu \rightarrow Cu.C_2 + H_2 \tag{5.2}$$

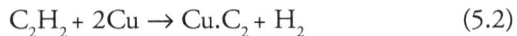

Surface Reaction

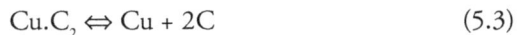

$$Cu.C_2 \Leftrightarrow Cu + 2C \tag{5.3}$$

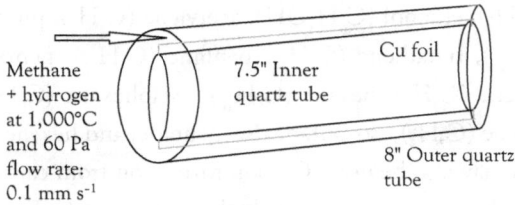

Figure 5.1 APFR for dissociation of methane and formation of graphene deposition on a copper foil

Given the monolayer formation of graphene, the Langmuir–Hinshelwood kinetics may be a reasonable assumption for the adsorption kinetics. At high temperatures such as 1,000°C, the dissociation rate can be expected to be rapid. Adsorption is a surface phenomenon. Molecules adsorb on the surface. Other molecules in the interior of the solid substrate are attracted by other surrounding molecules in all directions. The molecules that reside in the surface are in a state of imbalance and are pulled inward. The dissociated products from methane will interact with the molecules on the surface. The amount adsorbed would be proportional to the surface area available for adsorption. The Langmuir isotherm can be derived as follows:[3]

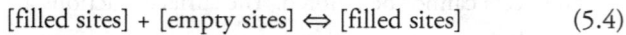

$$[\text{filled sites}] + [\text{empty sites}] \Leftrightarrow [\text{filled sites}] \qquad (5.4)$$

Sites are subject to chemical equilibrium:

$$[\text{bulk solute}] + [\text{empty site}] \Leftrightarrow [\text{filled site}] \qquad (5.5)$$

$$K' = \frac{[filled\_sites]}{[bul\_solute][empty\_site]} \qquad (5.6)$$

$$[filled\_sites] = K'[bulk\_solute]\frac{[total\_sites]}{1 + K'[bulk\_solute]} \qquad (5.7)$$

The rate of adsorption can be written as follows:

$$\text{Rate of adsorption "}r\text{"} = \frac{k_0 p_{C_2H_2}}{1 + K' p_{C_2H_2}} \qquad (5.8)$$

k_0 is a measure of the total concentration of sites available. Equation (5.8) is an alternative form of the Langmuir isotherm. The reciprocal of the rate of adsorption varies in a linear manner with the reciprocal of the partial pressure of acetylene. The equilibrium rate constant and the total concentration of the sites can be obtained from the slope and intercept of the straight line.

Acetylene decomposition can become autocatalytic.[4] It is a free-radical process. Free radicals can combine with the copper metal and become inert. This may keep the reaction from becoming a thermal auto-catalytic runaway. The axial flow in the annulus can be assumed to be in laminar flow. This is obtained from the Reynolds' number estimated for APFR of less than 50. As the reactant gas flows through the annulus, simultaneous reaction and diffusion can be expected. The dissociated products diffuse to the surface of the copper foil and get adsorbed. At high temperatures, the surface reactions of carbide formation and graphene formation are observed. The diffusion direction is radially inward. The cross-sectional area of the annulus is given by:

$$A_c = 0.25\pi \left(D_t^2 - D_c^2\right) \tag{5.9}$$

where D_t and D_c are the diameters of the outer tube made out of quartz and inner quartz tube wrapped with a copper foil. With the progress of CVD, the mole fraction of acetylene in the annulus decreases as the reactant flows down the length of the annulus. An effectiveness factor can be calculated to determine the overall rate of reaction per unit volume of the reactor space. The reactants diffuse radially inward toward the copper foil. Graphene deposits grow on the copper foil by the surface reaction given by Eqn. (5.3). Concentration of acetylene on the surface of the copper foil is less than the concentration of acetylene in the bulk. The effectiveness factor is defined as:

$$\eta = \frac{rate\_of\_reaction(actual)}{rate\_of\_reaction(if\_entire\_foil\_is\_at\_supplied\_concentration)} \tag{5.10}$$

$$\eta = \frac{2\int \pi D_c [-r"]dD}{\pi D_c^2 \left(-r"_{AA}\right)} = \frac{D_{C_2H_2}\left(\dfrac{\partial C_{C_2H_2}}{\partial r}\right)_{D=D_c}}{D_c\left(-r"_{AA}\right)} \tag{5.11}$$

Due to high temperature and low pressure, the primary mode of heat transfer is expected to be due to radiation. Small temperature differences may exist between the copper foil and GS. There is no need to couple the energy and mass balances at these small temperature gradients.

CVD methods can be used to prepare monolayer GS. Hydrocarbons are decomposed on transition metals such as nickel, copper, cobalt, and ruthenium. The hydrocarbons used are methane, ethylene, acetylene, and benzene.

Foils made of nickel and cobalt with thicknesses of 0.5 mm and 2 mm were used as catalysts.[5] These foils were chopped into five 5-mm² pieces and polished mechanically. The CVD process was carried out by effecting the decomposition of hydrocarbons around 800 to 1,000°C. Methane is passed at 60 to 70 sccm along with flow of hydrogen at 500 sccm at 1,000°C for 5 to 10 min over a nickel foil. Ethylene at 4 to 8 sccm can be used in place of methane at 60 to 70 sccm. Benzene can be used as the carbon source at 1,000°C for 5 min. Dilution of benzene is effected using argon and hydrogen. Acetylene at 4 sccm was decomposed on a cobalt foil at 800°C. Methane was decomposed at 1,000°C over a cobalt foil. The cooling step of the metal foils after the decomposition step was performed in a gradual manner.

Solvothermal Reduction Using N-methyl-2-pyrrolidone

A one-step, solvothermal reduction method for production of graphene oxide (GO) dispersions in *N*-methyl-2-pyrrolidone (NMP) is discussed by scientists at University of California, Los Angeles (UCLA).[6] A stable colloidal dispersion can be prepared by refluxing GO in NMP. Deoxygenation and reduction can be confirmed by a color change from brown to black. Single sheets of reduced GO (SRGO) were confirmed using scanning electron microscopy (SEM) and X-ray photoelectron spectroscopy (XPS). X-ray diffraction (XRD) analysis reveals a single broad peak at 3.4 A°. Stacking of GS can be confirmed.

Single-layer GO (SGO) is dispersed in dimethylformamide (DMF) with hydrazine hydrate.[7] Single-layer graphene (SG) is formed by reduction of SGO. Graphite oxide (GO) is prepared by oxidation of graphite by use of Hummers' procedure. An example of Hummers' procedure is as follows: 500 mg of graphite powder is reacted with 12 mL of concentrated sulfuric acid and 250 mg of sodium nitrate ($NaNO_3$) in a 500-mL flask under stirring and placed in an ice bath. Slow addition of 1.5 g potassium permanganate ($KMnO_4$) and temperature rise to 35°C are achieved. The time taken in this step is 30 min. In the next step, 22 mL of water is added slowly and the temperature is raised to 98°C. This is done for 15 min. Dilution of the reaction mixture is performed by addition of 66 mL of water. Three percent hydrogen peroxide (H_2O_2) treatment is done to decrease the consumption of water. The suspension formed is filtered in order to obtain a yellow-brown powder. Warm water is used to wash the product. The suspension is separated by ultrasonic treatment into SGO.

Ultrasonic treatment is carried out at 300 W and 33 kHZ. Yellow-brown powdery GO is recovered as a colloidal dispersion. Ultrasonic treatment at 35 KHz and 300-W power is used in order to produce SGO. Three hundred micrograms per milliliter SGO suspension is recovered from a water–N,N-dimethylformamide mixture and is treated with hydrazine hydrate at 80°C for 12 h. A black suspension of RGO in DMF–H_2O is formed.

Exfoliation Method Using a Carbonizing Catalyst

Graphene can be formed from a carbon source on a carbonizing catalyst. Large-sized GS can be prepared using this technique at a lower cost. GS may be damaged when exfoliated from the carbonizing catalyst after formation. A binding layer may be introduced in order to prevent damage of the GS. The carbonization catalyst may be selected from the following group in addition to copper: nickel, cobalt, iron, platinum, gold, silver, aluminium, chromium, magnesium, manganese, molybdenum, rhodium, silicon, tantallum, titanium, tungsten, uranium, vanadium, and zirconium. Neutrons should not be allowed to cause runaway reactions during the formation process. The carbonization catalyst film may either be thin or thick. The thin film has a thickness between 1 and 1,000 nm and the thick film has a thickness between 10 μm and 5 mm. In the section "Synthesis in an Annular Plug Flow Reactor and the Roll-to-Roll Transfer Process for Larger Areas," the CVD method was discussed.

Another method involves bringing in contact the carbonizing catalyst with a liquid-carbon-based material. The liquid-carbon-based material may be liberated by contact with the catalyst. The liberated carbon may become implanted in the catalyst. This process is called as carburization. The boiling point of the organic solution used may be 60 to 400°C. The liquid solution may be polar or nonpolar. It may be alcohol based, ether based, ketone based, ester based, organic acid based, and so on and so forth. These liquids may have different reduction abilities, reactivities, and adsorption rates. The heat treatment may be performed under well-stirred conditions at a temperature range of 100 to 400°C. The contact time may range from 10 min to 48 h. A binder layer and plastic substrate, such as a polyethylene terephthalate (PET) layer, are formed over the graphene layer. The formed GS may be transferred to a target device such as a barrister using dry and wet etching methods. The GS is exfoliated by Samsung[8] using acid solutions such as sulfuric acid, nitric acid, and hydrochloric acid. The binder layer may be made out of an insulating material such as siloxane based, acryl based such as PMMA, or epoxy based such as an epichlorohydrin compound. A photoresist material or a polymer electrode material, such as polyphosphogen, may also be used as the binder layer.

Ion Implantation Method for Layer Thickness Control

Scientists at Harvard University[9] have developed an ion implantation method for graphene synthesis with layer-by-layer thickness control. Precise doses of carbon atoms are introduced into polycrystalline nickel films. Heat treatment, as a subsequent step, may be used to grow them on the nickel film surface. There is some solubility of carbon atoms in nickel at 1,000°C. The bulk solubility of carbon in nickel is reduced at lower temperatures. This is used to cause segregation of carbon atoms on the surface of nickel and lead to the formation of graphene layers by the phenomenon of crystallization. Layer-by-layer thickness control is achieved using 15% less carbon used in the CVD process. A 40-nm penetration depth of carbon ions is achieved. Almost all of the carbon atoms implanted crystallized into graphene. This can lead to more precise control of layer thickness. The process comprises the following steps:

1. Evaporation of nickel and deposition onto Si/SiO$_2$ wafers in order to attain 500-nm thickness;
2. Annealing in argon and hydrogen at 1,000°C at ambient pressure for 2 h;
3. Recrystallization of the nickel film into 2-μm average-size grains. XRD analysis revealed
4. Miller plane (111) oriented parallel to the film thickness as expected;
5. Dosing of nickel with carbon ions with 30 keV at 2 to 13 Peta ions cm^{-2}. The number of layers of graphene formed depends on the dosage level;
6. Heat treatment at 1,000°C under vacuum pressure of under 4 Pa for 1 h. This step allows for bulk diffusion of carbon atoms in nickel. The sample is mounted on a button heater;
7. Cooling of a supercritical solid solution to room temperature at heat removal rates of 5 to 20°C min^{-1};
8. Crystallization of dissolved carbon atoms into graphene;
9. Transfer of graphene into a Si/SiO$_2$ chip using the PMMA transfer method; and
10. Unique characteristics of graphene confirmed using Raman spectroscopy.

Two-dimensional resistivity of 2 kΩ per square has been reported for GS. In addition to the film thickness, other properties are also affected by the ion dosage level. The diffusion of carbon ions in nickel during the growth of graphene layers can be modeled as follows:

The Stokes–Einstein equation can be written as:

$$\begin{pmatrix} ion \\ velocity \end{pmatrix} = \begin{pmatrix} ion \\ mobility \end{pmatrix} \left(\Sigma\ forces \right) \qquad (5.12)$$

Two kinds of forces can be expected, a force due to variation in chemical potential and the other due to electrical forces. The expression for ion mobility is given by including an acceleration term[10] to the expression by Cussler[11] and can be obtained as:

$$\left(nM \frac{\partial v_{CNi}}{\partial t} + 3\pi\eta \, d_0 v_{CNi} \right) = -\nabla\mu_{CNi} - z_C \nabla\psi \qquad (5.13)$$

where u_i is ion mobility, z_c is the ionic charge (+1 for carbon), F is the Faraday's constant, ψ is the electrostatic potential, n is the number of moles of carbon, and M is the molecular weight of carbon. The acceleration term[12] is taken into account in Eqn. (5.13). The ion mobility is assumed to be in the Stokes settling regime and taken as:

$$u_i = \left(\frac{1}{3\pi\eta \, d_0} \right) \qquad (5.14)$$

where d_0 is the diameter of the ion and η is the viscosity of the medium. The use of this equation for ion mobility in solid solutions needs to be confirmed experimentally. Assuming an ideal solution, the activity coefficient of carbon in nickel may be assumed to be 1. The contribution from the gradient in chemical potential can be written as:

$$\nabla\mu_{CNi} = \frac{RT\nabla C_{CNi}}{C_{CNi}} \qquad (5.15)$$

Plugging Eqn. (5.15) in Eqn. (5.13) and realizing that the flux of ions $-J_C = C_{CNi} v_{CNi}$

$$\left(\tau_r \frac{\partial J_{CNi}}{\partial t} + J_{CNi} \right) = -D\nabla C_{CNi} - \frac{C_{CNi} z_c F \nabla\psi}{RT} \qquad (5.16)$$

where τ_r is the relaxation time of carbon atoms on the nickel surface. Equation (5.16) is a modified Nernst–Planck equation written by Bard and Faulkner.[13,14] Modification is to include the acceleration effects of the carbon ion during transient diffusion.

Chemical Method

Graphene can be synthesized from ethyl alcohol in a medium pressure autoclave with a stainless steel chamber for 20 h at 230°C and 60 bar. Eighty milliliters of C_6H_5OH was mixed[15] with the solution of 4.3 g of $NaBH_4$ and 15 mL of 10 M NaOH. The reaction was carried out in the presence and absence of sodium dodecyl sulfate (SDS) at 2 wt% concentration. The reaction product obtained after filtering was dried in a hot oven for 10 h at 60°C. The samples were characterized using SEM, TEM, powder XRD, and AFM. Ethanol was found to react with $NaBH_4$ and H_2, alkyl borates, and alkoxy borohydrides were produced.

High concentration of aqueous NaOH was found to be essential to form graphene layers. Reactions that were performed at lower temperatures were found to show lesser yield. Ethanol acts as a carbon source as well as a solvent. This process is less expensive and has potential for industrial scalability. This method is similar to the Wolff–Kishner reduction process to prepare multiwalled carbon nanotubes (MWCNTs). A surfactant is used in place of post-pyrolysis of the product obtained in order to achieve the improved yield and dispersion of graphene.

Large-Area Synthesis by Phase Separation of Carbon-Metal-Miscible Melts

A scalable, high-throughput production technological route to large-area GS was developed by scientists at the University of California, Riverside.[16] Carbon is dissolved in molten transition metals such as nickel and copper at a specified temperature. Later, the dissolved carbon is allowed to nucleate and grow atop the melt at a cooler temperature. Raman two-dimensional bands were used to determine the number of atomic planes in the resulting graphene layers. High-quality SG on metals is produced and interface materials are used for thermal management applications. Nickel and a graphite source are allowed to melt in a crucible. When the temperature is lowered, thermiscible solids will undergo phase separation. Excess carbon crystallizes into graphene atop the melt. The time–temperature diagram looks like an Olympic medal podium with T_2 where the gold medalist stands when receiving the medal, T_m

where the silver medalist stands, and T_3 where the bronze medalist stands.

Electric resistance heating at 75 A is used for the heating process for 20 s. The furnace is operated under vacuum at 77 Pa. A hypereutectic composition of nickel + 2.35 wt% carbon was selected. The temperatures T_2, T_m, and T_3 are shown in Figure 5.2. The metal substrate is dissolved, and graphene layers are transferred to a silicon wafer. To serve this end, a layer of PMMA was spin coated on the substrate at 1,800 rpm for 30 s. The metal substrate was etched by a nitric acid solution. Transfer was accomplished using the steps of washing with isopropanol and water and annealing at 60°C for 1 h. Film morphology with wrinkles separating smooth regions was found.

Unzipping CNTs by Chemical–Thermal Method

There are two chemical–thermal processes to form graphenes from CNTs reported by scientists at Richard E. Smalley Institute for Nanoscale Science and Technology, Rice University, Houston, TX. The first is a two-stage[17] procedure comprising of an oxidation step and a reduction step. A suspension of MWCNTs in concentrated sulfuric acid (H_2SO_4) is treated with excess potassium permanganate. The acidic conditions are needed to exfoliate the GS nanostructure. The reaction mixture is well stirred for 1 h at room temperature and heated to 50 to 70°C for an additional hour. The reaction mixture was quenched by pouring over ice containing a small amount of hydrogen peroxide. The solution is filtered over a polytetrafluoroethylene (PTFE) membrane and the remaining solid was washed with acidic water followed by ethanol. The second stage is to reduce oxidized nanoribbons into graphene. An aqueous solution of nanoribbons was treated with 1 wt% SDS, surfactant, and 1 vol% hydrazine monohydrate (N_2H_4–H_2O). The solution was covered with a thin layer of silicon oil before being heated to 95°C for 1 h.

An alternative[18] for preparation of graphene by unzipping of CNTs was presented by a reaction of MWCNTs with potassium. Potassium was melted over MWCNTs under a vacuum pressure of 6.7 Pa. One gram of MWCNTs and 3 g of potassium pieces were placed in a 50-mL pyrex ampule that was evacuated and sealed with a torch. The reaction was

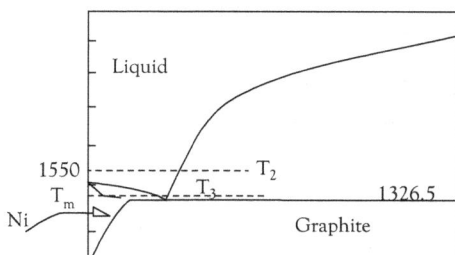

Figure 5.2 Phase diagram of a nickel–carbon system

carried out in a furnace for 14 h at 250°C. The heated ampule containing a golden-bronze-colored potassium intercalation compound and silvery droplets of unreacted metal was cooled to room temperature. The ampule was opened in a dry box in a nitrogen-filled glove bag and then mixed with 20 mL ethyl ether. Twenty milliliters of ethanol was slowly added into the mixture of ethyl ether and potassium-intercalated MWCNTs at room temperature. Heat is dissipated by evolution of hydrogen gas bubbles. The quenched product was removed from nitrogen enclosure and collected on a PTFE, 450 nm. Then the sample was washed with 20 mL of ethanol, 20 mL of water, 10 mL of ethanol, and 30 mL of ether. Next, the sample was dried under vacuum in order to produce longitudinally split MWCNTs. The external appearance was a 1 g black, fibrillar powder. Chlorosulfonic acid was used to exfoliate potassium-split MWCNTs. An ultrasonic jewelry cleaner was used to make a dispersion of MWCNTs in chlorosulfonic acid by sonication for 24 h. The mixture was quenched by pouring the solution onto ice followed by filtration using a PTFE membrane. The filter cake was dried under vacuum. A stock solution of graphene was obtained by dispersion of a black powdery material in DMF.

A team from Stanford University developed a method where the nanotubes are embedded in the polymer film and the tubes are etched with an argon plasma. The film is then removed using solvent and application of heat. The SWCNTs and MWCNTs are unzipped and the graphene ribbons are formed that are 10 to 20 nm wide. This procedure was found to work well with MWCNTs with lots of surface defects. Entanglement of unzipped tubes was a concern. Research is underway to untangle the ribbons. The control of width and edge patterns of nanoribbons needs more work.

Nanoribbon Alternation from Flakes

Alternating layers of graphene with nanoribbon morphology and insulation layers made up of graphane or partially hydrogenated graphene were patented by IBM, Armonk, New York.[19] This kind of nanostructure (Figure 5.3) is needed in the electronics industry. This can be used as an alternative to silicon semiconductors and make good field-effect transistors (FETs). A gate dielectric layer is formed on top of the nanoribbon-containing layer. Source and drain regions are positioned on opposite sides of the conductor.

The trends in the semiconductor electronics industry include miniaturization of devices that get fabricated, devices that use less power than the previous generation of devices, and devices that operate faster. Personal devices such as cell phones and personal computers, both desktop and laptop, which emerge in the markets are made more portable. These devices need increased memory and more computational speed and power. There is an expectation in the industry for smaller and faster transistors with speeds greater than 30 pHZ. This limit was estimated based on the current laws of physics in collegiate textbooks and a limit on gate width of 10 to 15 nm and speed of light. Only an atomic layer in thinness, graphene has the potential for microprocessor speed increases to 23 EHz. The delocalized electrons in the p orbital in the polyatomic hexagonal graphene rings can form an n junction in transistor devices. Currently, monolayer graphene has been found without a bandgap. Smaller separations between npn junctions are needed for further speed increases. Nanoribbon morphology can be used to create the bandgap that can make a difference for better speed increases. The graphene nanoribbon (GNR) width has to be less than 5 nm for significant increase in the bandgap. The size of the bandgap increases with decreasing width of nanoribbons. Alternation of layers is accomplished by a reactive ion etching (RIE) method.

Electrophoretic Deposition and Reduction

A graphene film is generated by in situ electrochemical reduction of a GO film. Prior to this step, the GO film is formed on the ITO substrate by electrophoretic deposition. The maximum specific capacitance of the prepared GS film electrode was calculated to be 156 F g^{-1}.[20]

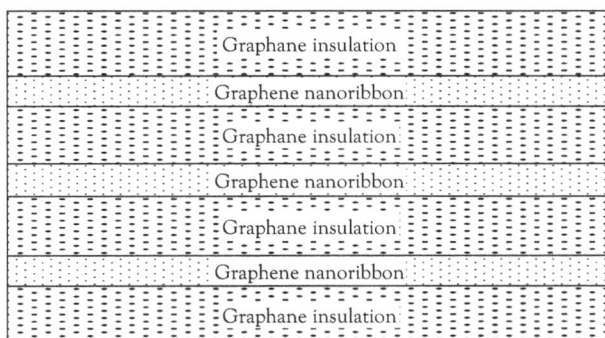

Figure 5.3 Alternate-layer morphology of graphene and graphane

The capacitance retention of the material was 78% upon 400 times of cycling. XPS measurements were used to confirm the removal of oxygen-containing functional groups in a GO film. Potentially, this system can be used to prepare supercapacitors. Graphene-based ultracapacitors can be used to store more energy, have longer life, be lighter in weight, be more flexible, and be easier to maintain. They hold promise for power energy storage applications. Carbon-based materials such as CNTs, GS, carbon fibers, activated carbon are contemplated for making the third-generation supercapacitors. They exhibit higher capacitance compared with conventional capacitors. GS has a high surface area, that is, about 2,630 m^2 g^{-1}. They also possess superior mechanical stiffness, flexibility, and electrical properties. The electrophoretic deposition method is better compared with other methods such as solvent dispersion that result in poor dispersion of GS and poor adhesion to the substrate.

A stainless steel plate and a clean ITO conductive glass sheet were used as anode and cathode respectively. An applied electromotive force (EMF) of 150 V was provided for 45 s in order to allow deposition of GO films on the ITO cathode. The GO aqueous suspension had a concentration of 0.6 mg mL^{-1} at room temperature. The electrochemical reduction was accomplished using a CHI660C electrochemical workstation in the 0.1 M KCl aqueous solution. Surface morphology and microstructure were characterized using field–emission SEM (FE-SEM), XPS, and electrochemical impedance spectroscopy (EIS). A three-electrode system was in operation in the electrochemical workstation.

These electrodes are (1) ITO anode where the GS film is deposited form the working electrode; (2) platinum wire as counter electrode; and (3) Ag/AgCl was used as the reference electrode. The electrolyte used was a 0.1 M sodium sulfate (Na_2SO_4) solution. Utrasonification was used in order to form the well-dispersed, brown-colored GO aqueous solution.

Flash Cooling

Some of the methods used to prepare graphene in larger areas and with higher quality require explosive gases such as methane, need be operated under vacuum, need high-cost instrumentation, and need a high level of alarms and safety precautions. Facile synthesis of high-quality SG from ethanol using "flash cooling" right after the CVD step was reported by Miyata et al.[21] Graphenes were synthesized by allowing for the reaction of ethanol with nickel substrate at 900°C. A 3-cm-diameter quartz tube of 100 cm length was filled with argon gas at a flow rate of 300 cc min[-1]. Gas flow is allowed to continue through the synthesis and cooling steps. Nickel foils with 5-μm thickness were inserted into a preheated quartz tube at 600°C. The temperatures are allowed to rise to 900°C. Ethanol is bubbled using argon gas at 200 cc min[-1]. The bubbling was stopped after 5 min of reaction. Nickel substrates were cooled at different rates. During the flash cooling step, the substrate was immediately removed from the quartz tube and then cooled from 900°C to 560°C. The time period of the cooling step was 10 s. Natural cooling took 20 min for the same decrease in temperature. Graphene formed on reaction with nickel was transferred into a SiO_2/Si wafer using nitric acid. The procedure was conducted at atmospheric pressure. It is believed that SG growth does not occur during the carbon precipitation. The SG stems from a phenomenon called "surface diffusion" of carbon on a nickel substrate.

Flash cooling was found to be a necessary step subsequent to the CVD step in order to obtain high-quality SG. Comparison studies were conducted using naturally cooled and flash cooled substrates. The scientists concluded that decomposition and formation of graphenes occur rapidly after the supply of the carbon source is stopped. Ethanol decomposition into graphenes on a nickel substrate was confirmed by "reannealing experiments." Prior to the experimentation, the presence of graphenes was confirmed using Raman spectroscopy at room temperature. Carbon first dissolves into the nickel substrate. Concomitantly, carbon precipitates into graphene. Monolayer and multiple-layer formation depends on the cooling rates. At 1,000°C, the solubility of carbon in nickel is high. As the temperature was decreased, the solubility decreased. A supersaturated solution upon perturbation can form crystals. The graphene allotrope is formed upon cooling. Typical cooling rates deployed are around 10°C s^{-1}. Quenching was accomplished at extremely fast cooling rates. In the case of use of copper substrates, low carbon solubility and surface catalytic activity combined effects lead to graphene allotrope formation.

Carbon solubility in bulk nickel is usually high. Formation of nickel oxide on the surface of nickel leads to reduction in carbon solubility. This may have contributed to the growth of the graphene layer. The surface diffusion phenomenon is suspected to play an important role in the formation of graphene monolayers. This suspicion was confirmed by an effect of reaction time. A thirty-minute reaction time led to the formation of multiple layers of graphene. Carbon dissolution may accelerate the segregation of multilayer graphenes. Etching of nickel oxide can result in promotion of the dissolution process of carbon.

Gas Intercalation and Exfoliation

CNTS are higher in cost due to the poor yield and low production and purification rates using the current methods of production. Nanoscaled graphene platelets can potentially be a cost-effective substitute for CNTs. Some of the processes used currently need costly waste disposal from the washing step. High furnace temperatures of 500 to 2,500°C are used in some processes. A lower-cost method was patented by Nanotek Instruments, Dayton, OH.[22]

Graphene platelets contain one to five layers of GS with each layer having a thickness of 3.4 A°. For graphite oxide flakes, each sheet is about 6.4 to 10.2 A°. Production of these materials is accomplished in three steps: (1) pre-pressurization; (2) intercalation; and (3) exfoliation. The starting material such as graphite sized about 10 μm is brought in contact with high-pressure gas at pressures of 2 to 10 atm and temperatures of 50 to 200°C. The gas species is made to intercalate into the interstitial and interlayer spaces in the layered material. The material is charged into the intercalation chamber (Figure 5.4). A nozzle is used to discharge the intercalated mixture into the exfoliation chamber. The exfoliation chamber is operated at a different pressure and temperature. Here, the nanosized graphene platelets are exfoliated from the mixture. The gas can be selected from hydrogen, helium, neon, argon, nitrogen, oxygen, fluorine, and carbon dioxide. The layered starting material can be selected from graphite, graphite oxide, graphite fluoride pre-intercalated graphite, graphite or carbon fibers, graphite nano-fibers, clay, bismuth selenides or tellurides, transition metal dichalcogenides, sulfides, selenides or tellurides of niobium, molybdenum, halfnium, tantalum, tungsten, or rhenium. The pre-pressurizing step is conducted at room temperature and the starting material is brought in contact with the pressurized gas. The gas supply to the intercalation chamber is from a gas cylinder as shown in Figure 5.4.

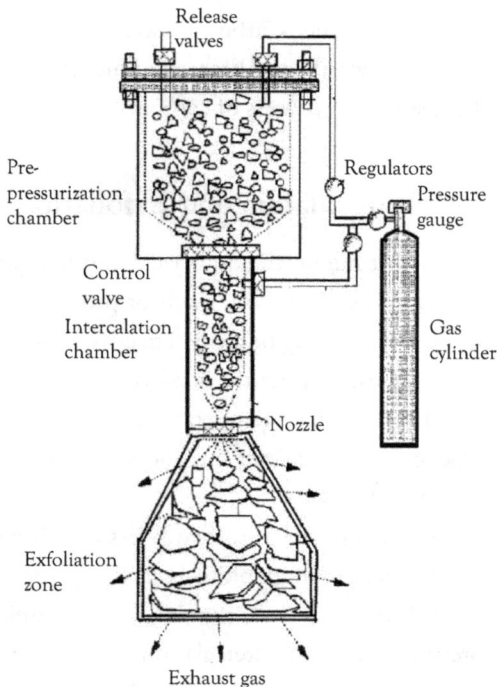

Figure 5.4 Pre-pressurization, intercalation, and exfoliation stages during production of graphene nanoplatelets

Pressure regulators are used to control the pressure in the exfoliation and intercalation chambers. The solubility of the gas in the solid increases with increasing pressure. The gas intercalation time can range from a few minutes to a few hours. A gas release valve is used to remove excess gases. The use of gas for intercalation can result in a lower cost of production.

Exfoliation is achieved by reduction of pressure rather than by use of high temperature. This is a semi-batch process. Use of environmentally benign gases and lower temperatures can result in lower cost of production in large scale of GS. Fully isolated and separated platelets are found from this process. A nozzle is used to rapidly propel the gas intercalated material from the intercalation chamber into the exfoliation chamber. Exfoliation is instantaneous. The exfoliation chamber is made with perforations in order to allow for exhaust gases to exit the chamber.

The diffusion time of gases, $\tau = \left(\dfrac{\lambda^2}{D_{AB}} \right)$, where λ is the diffusion path and D_{AB} is the binary diffusion coefficient ($cm^2\ s^{-1}$) in a 100-μm graphite flake, can be estimated to be about 11.5 days. This would imply a large intercalation time. By increasing the temperature and/or by reduction of the flake size, the intercalation times can be reduced. The diffusion coefficient of the intercalant in the graphite flake doubles every 10°C or increases in an Arrhenius manner. When the flake size is reduced from 100 μm to 1 μm, the diffusion time will reduce to 100 s at room temperature. This can be further reduced by increasing the intercalation chamber temperature. These estimates of diffusion times can be refined by use of damped wave diffusion and relaxation models.[23] In these cases, Knudsen diffusion effects may be significant. For materials with large relaxation times, the diffusion process is characterized by the damped wave diffusion and relaxation mechanism. The centerline transient concentration for the case of a finite slab subject to a step change in the concentration can be given as[24] follows:

$$u = \sum_{1}^{\infty} c_n e^{-\frac{\tau}{2}} Cos\tau \left(\sqrt{\lambda_n^2 - 0.25} \right) \qquad (5.17)$$

where u is the dimensionless concentration, $u = \left(\dfrac{C_A - C_{As}}{C_A - C_{Ai}} \right)$, C_{Ai} is the initial concentration of the gas in the finite slab of width $2a$, C_{As} is the concentration of the gas at the surface, C_A is the concentration of the gas in the slab as a function of space and time, τ is the dimensionless time and given by $\left(\dfrac{t}{\tau_r}\right)$, τ_r is the relaxation time of the material. It was shown in the work by Sharma (2005) that the values of the constants in the infinite series in Eqn. (5.17), c_n and the eigenvalues, λ_n was given by:

$$c_n = \left(\frac{4(-1)^{n+1}}{(2n-1)\pi}\right); \ \lambda_n = \left(\frac{\sqrt{D_{AB}\tau_r}\,(2n-1)}{2a}\right), \ n = 1, 2, 3, \cdots \quad (5.18)$$

It can be realized from Eqn. (5.17) that a steady-state concentration will be attained after a said time t_{ss}. At this point, the value of u would become zero. This is one of the novel features of the damped wave diffusion and relaxation model for materials with large relaxation times:

$$\tau_r > \left(\frac{a^2}{\pi D_{AB}}\right) \quad (5.19)$$

At the point where the concentration reaches steady state, the value of the argument in cosine function in Eqn. (5.17) would become $\pi/2$. At this juncture,

$$\frac{\pi}{2} = \left(\frac{t_{ss}}{\tau_r}\right)\sqrt{\lambda_1^2 - 0.25} = \left(\frac{t_{ss}}{\tau_r}\right)\sqrt{\frac{D_{AB}\tau_r}{4a^2} - 0.25} \quad (5.20)$$

or

$$t_{ss} = \left(\frac{\tau_r}{\sqrt{\dfrac{D_{AB}\tau_r}{4\pi^2 a^2} - 0.02533}}\right) \quad (5.21)$$

Equation (5.21) is applicable for materials with large relaxation times as given by Eqn. (5.19). Equation (5.19) is applied to the 1-μm graphite flake and the time to steady state after the transient diffusion of gas process for intercalation is completed can be estimated at about 22.2 μs. The relaxation time was estimated from acceleration of the molecule considerations as follows:[25]

$$\tau_r = \left(\frac{k}{PC_p}\right) \quad (5.22)$$

The thermal conductivity for graphene is taken as 13 times that of copper and is about 5,248 W m^{-1} K^{-1}, the pressure is taken as 20 atm in the intercalation chamber, and heat capacity is taken as $5R/2$. For the 100-μm flake, as per Eqn. (5.19), damped wave effects will not be dominant and the Fick diffusion model is sufficient to describe the transient diffusion events.

The exfoliation step follows the intercalation stage (Figure 5.4). The gas-intercalated material is discharged through a nozzle into the exfoliation chamber. The reduction in pressure in the exfoliated chamber results in reduction in the solubility of the gas in the graphite material. The gas–solid system can be expected to be in a supersaturated state. Gas species that is in excess can be expected to exit the solids. Due to the driving force of the pressure difference between the interstices of the solid material and the exfoliation chamber, the gas species can be expected to overcome the van der Waals forces that keep the graphene planes together. Layer separation of the graphene flakes seems to occur in this manner. In summary, when the laminar material is subjected to high pressure gas, the gas molecules penetrate into the interstitial space of the solid up to a point where the internal pressure will equal the intercalation chamber pressure. When the solid material is subjected to a lower pressure as is the case in the exfoliation chamber, the gases in the interstitial spaces will have a tendency to expand. This inclination is sufficient to overcome the intersheet van der Waals forces and thus effect layer separation of the graphene layers. Subsequent to exfoliation, a mechanical attrition step may be included in order to achieve the desired particle size reduction. Grinding, pulverization, ultrasonication, and milling are examples of mechanical attrition methods. Ball-milling is an effective method for mass production at lower costs. Supercritical carbon dioxide can be used as the intercalating gas that is later exfoliated.

Graphene Shell Formation

In the micromechanical method, a SCOTCH™ tape from 3M Corporation, Minneapolis, MN, is attached to a graphite sample. When the tape is detached, the graphene layer is separated with it. In this case, the separated GS does not have uniform number of layers and it is not clear how large-area GS can be made using this method. Samsung[26] patented a lower-cost process to make graphene in the form of a shell structure.

A polymer that is hydrophobic–hydrophilic is coated on a catalyst that can be used in graphitization. Heat treatment results in the formation of graphene shells. The shells can be made outside solid spheres, cylinders, polyhedrons, and plates. The shape imparted to the graphene shell can be controlled by selection of shape of the catalyst used. The catalyst that can be used is from the class mentioned in the section "Large-Area Synthesis by Phase Separation of Carbon-Metal-Miscible Melts."

Amphiphilic polymer, liquid crystal polymer (LCP), or conductive polymer can be used for the coating. The hydrophilic group is that of hydroxyl, carboxyl, sulfate, sulfonate, phosphate, or phosphonate functional groups. Examples of hydrophobic groups are those of halogen, halogenated alkynyl, alkenyl, alkoxy, hetroalkyl, aryl, and arylalkyl functional groups. Polyacetylene and polythiopene are examples of conductive polymers. The polymer may be formed in situ during the coating step. The time and temperature needed for the heat-treatment step is about 400 to 2,000°C and 6 to 600 min. Upon heat treatment, the graphene layer is separated by the catalyst by acid treatment. The thickness of the polymer coating can be used to control the thickness of graphene layers. The thickness of the graphene shell can range from 1 A° to 100 nm. The diameter of the shell is made greater than 1 μm. Thermal energy can be provided by induction heating, laser irradiation, infrared (IR) irradiation, microwave irradiation, plasma heating, ultraviolet (UV) irradiation, surface plasmon heating, and so forth.

Coal Tar Pitch as a Source

Coal tar pitch may be used as a starting material for preparation of graphenes. Heat treatment of coal tar pitch followed by the steps of exfoliation and mechanical attrition can lead to nanoscale graphene plate (NGP) materials. The heat-treatment temperature may range from 300°C to 1,000°C for partial carbonization. Complete carbonization may need heat treatment in the temperature range of 1,000 to 3,000°C. The electrical, mechanical, and thermal properties of NGP materials are found to be comparable to CNTs. The GS is made by longitudinal scission of CNTs.[27] Coal tar pitch or polyacrylonitrile (PAN) polymer can be used as the starting material. A polymeric carbon with high aromaticity is formed using the heat-treatment step. Nanometer-scaled crystallites are formed. A GS is a two-dimensional crystalline material. Subsequent to heat treatment, the crystallites need to be exfoliated. Delamination of the GS is accomplished in the exfoliation step. Exfoliation may be achieved using chemical treatment, intercalation, foaming, or heating and cooling steps. The third step is attainment of desired size reduction by mechanical attrition.

Single-layer NGPs or materials with stacks of graphene planes are reduced to sizes smaller than 10 nm. In the c axis, some graphene planes may remain bonded to each other by weak van der Waals forces. During the carbonizing heat-treatment step, the non-carbon portions of the materials exit the material. The remaining solid is polymeric carbon. Polymeric carbons can exist in amorphous or crystalline forms. A spectrum of materials with varying proportions of amorphous and crystalline forms and defects is possible. Prior to formation of graphene, polyacene with repeat units of (C_4H_2) may have formed. These materials later can be transformed into wider aromatic ring structures as in GS. They can form stacks and thicker plates at higher heat-treatment temperature and times and can form a turbostatic structure in carbon fibers.

Fullerenes

Fullerenes, C_{60}, are the third allotropic form of carbon. The Nobel Prize for their discovery was awarded in 1996 to Curl, Smalley, and Kroto. The soccer-ball structure of fullerenes, C_{60}, with a surface filled with hexagons and pentagons satisfies Euler's law. Euler's law states that no sheet of hexagons will close. Pentagons have to be introduced for hexagon sheets to close. Stability of C_{60} requires Euler's 12-pentagon closure principle and the chemical stability conferred by pentagon nonadjacency. C_{240}, C_{540}, C_{960}, and $C_{1,500}$ can be built with icosahedral symmetry.

Howard patented the first-generation combustion synthesis method[28] for fullerene production, an advancement over the carbon arc method. The second-generation combustion synthesis method optimizes the conditions for fullerene formation. A hydrocarbon with a continuous high flow is burnt at low pressure in a three-dimensional chamber. Manufacturing plants have been constructed in Japan and the United States with a production capacity of fullerenes at 40 metric tons y^{-1}. Purity levels are greater than 98%. The reaction chamber consists of a primary zone where the initial phase of combustion synthesis is conducted and a secondary zone where combustion products with higher exit age distribution do not mix with those with lower exit age distribution. Flame control and flame stability are critical in achieving higher throughputs of fullerenes. Typical operating parameters include residence time in the primary zone of 2 to 500 ms, residence time in the secondary zone of 5ms to 10 s, total equivalence ratio in the range of 1.8 to 4.0, pressure in the range of 10 to 400 Torr, and temperature in the range of 1,500 to 2,500 K.

Fullerene crystals can be produced with high yield. By counterdiffusion from a fullerene solution to a pure isopropyl alcohol solvent, fullerene single-crystal fibers with needle shape were formed. Needle diameters were found to be 2 to 100 μm and lengths were 0.15 to 5 mm. Buckyball-based sintered carbon materials can be transformed into polycrystalline diamonds under less severe conditions using powder metallurgy methods.

A chemical route has been developed by Scott to synthesize C_{60}. Corannulene is synthesized from naphthalene. As the rings fuse and the sheet forms, it is rolled into a soccer-ball structure. The challenge is how to stitch up the seams between the arms to make the ball. Oligoarenes are transformed into highly strained curved Pi surfaces. The molecule needs to bend to effect ring closure in a "soccer-ball" structure at 1,000°C. A 60-carbon ring system can be built by acid-catalyzed aldol trimerization of ketone. Oligoarene zips up to the soccer-ball structure effected by cyclodehydrogenations.

In order to generate higher yield, supercritical ethanol was used to react with naphthalene with ferric chloride as the catalyst for 6 h. The reaction products were subjected to extraction with toluene. The reactor temperature range was 31 to 500°C and pressure range was 3.8 to 60 MPa. Smalley patented a process to make fullerenes by tapping into the solar energy. The carbon is vaporized by applying a focus of solar arrays and conducting the carbon vapor to a dark zone for fullerene growth and annealing. The fullerene content of soot deposits collected on the inside of the pyrex tube was analyzed by extraction with toluene.

In the electric arc process for fullerene production, a carbon material is heated using an electric arc between two electrodes to form carbon vapors.

Fullerene molecules are condensed later and collected as soot. Fullerenes are later purified by extraction of soot using a suitable solvent followed by evaporation of the solvent to yield the solid fullerene molecules.

Applications of fullerenes include higher-temperature superconductors. Polymerized fullerene molecules were found to have a transition temperature of 100 to 150 K; adsorbents with improved gas storage capability; excellent catalysts such as bucky anions to convert ethyl benzene to styrene for example; fullerene composite with improved mechanical strength; advanced electromechanical systems where a proton conductor is sandwiched between two porous electrodes; and synthetic diamonds by static crystallization and shock conversion methods.

Carbon Nanotubes

CNTs are rolled GS of atoms about their needle axis. They are 0.7 to 100 nm in diameter and a few microns in length. Carbon hexagons are arranged in a concentric manner with both ends of the tube capped by pentagon-containing Buckminster fullerene-type structure. They possess excellent electrical, thermal, and toughness properties. Young's modulus of a CNT has been estimated as 1 TPa with a yield strength of 120 GPa. S. Ijima verified fullerene in 1991 and observed MWCNTs formed from carbon arc discharge.

Five methods of synthesis of CNTs are discussed. These are:

1. Arc discharge;
2. Laser ablation;
3. CVD;
4. High-pressure carbon monoxide (HIPCO) process; and
5. Surface-mediated growth of vertically aligned tubes.

The arc discharge process was developed by NEC in 1992. Two graphite rods are connected to a power supply spaced a few millimeters apart. At 100 A, carbon vaporizes and forms a hot plasma. The typical yield values are 30 to 90%. The SWCNTs and MWCNTs are short tubes 0.6 to 1.4 nm in diameter. It can be synthesized in open air. The product needs purification. The CVD process was invented by Nagano, Japan. The substrate is placed in an oven, heated to 600°C, and a carbon-bearing gas such as methane is slowly added. As the gas decomposes, it frees up the carbon atoms that recombine as a nanotube. The yield range is 20 to 100%. Long tubes with diameters ranging from 0.6 nm to 4 nm were formed. They can be easily scaled up to industrial production. The SWCNT diameter is controllable. The tubes are usually multiwalled and riddled with defects. The laser vaporization process was developed by Smalley in 1996. The graphite is blasted with intense laser pulses to generate carbon gas. A prodigious amount of SWCNTs is formed. A yield of upto 70% is found. Long bundles of tubes 5 to 20 μm in length with diameters in the range of 1 to 2 nm are formed. The product formed is primarily SWCNTs. Good diameter control is possible and few defects are found in the product. The reaction product is pure. The process is expensive.

The HIPCO process was also developed by Smalley in 1998. A gaseous catalyst precursor is rapidly mixed with carbon monoxide in a chamber at high pressure and temperature. The catalyst precursor decomposes and nanoscale metal particles form the decomposition product. Carbon monoxide reacts on the catalyst surface and forms solid carbon and gaseous carbon dioxide. The carbon atoms roll up into CNTs. Hundred percent of the product is SWCNTs and the process is highly selective. Samsung patented a method for vertically aligning CNTs on a substrate. A CNT support layer is stacked on the substrate filled with pores. A self-assembled monolayer (SAM) is arranged on the surface of the substrate. To the end of each of the CNTs are attached portions of the SAM exposed through the pores formed between the colloid particles present in the support layer. CNTs can be vertically aligned on the substrate having the SAM on it with the help of pores formed between the colloid particles.

CNTs possess interesting physical properties. The thermal conductivity of CNTs is in excess of 2,000 w m^{-1} K^{-1}. They have unique electronic properties. Applications include electromagnetic shielding, electron field emission displays for computers and other high-tech devices, photovoltaics, supercapacitors, batteries, fuel cells, computer memory, carbon electrodes, carbon foams, actuators, materials for hydrogen storage, and adsorbents.

CNTs can be produced with different morphologies. Examples of different morphologies include SWCNT, DWNT, MWCNT, nanoribbon, nanosheet, nanopeapod, linear and branched CNT, conically overlapping bamboo-like tubule, branched Y-shaped tubule, nanorope, nanowire, and nanofilm. Processes are developed to prepare CNTs with desired morphology. Phase-separated copolymers/stabilized blends of polymers can be pyrolyzed along with a sacrificial material to form the desired morphology. The sacrificial material is changed to control the morphology of the product. Self-assembly of block copolymers can lead to 20 different complex phase-separated morphologies. Oftentimes, as is the precursor so is the product. Therefore, CNTs with even more variety of morphologies can be synthesized.

Electrochemical Method

An electrochemical route to few-layer graphene flakes has been developed at the University of Texas, Austin, TX, and at the National University of Singapore. The route is based on electrode swelling noted during battery charging. This swelling was detrimental to the battery operation. The graphene flake synthesis is made possible by tuning the parameters that go into swelling of the electrode during battery charging. A graphite electrode is charged in an electrolyte solution of lithium salts and propylene carbonate at 15 V. The product was sonicated. Lithium ions and decomposition products of the solvent are forced to rapidly intercalate into graphite and the material is exfoliated. Graphene ink prepared using this method can be used in order to make electrically conducting papers. This is another method to make graphene from graphite down to a few atomic layers in thickness. Graphite electrodes are routinely used in industries and this route to graphene is scalable.

Zinc Acid Process

Chemists at Rice University have found a method for etching a multilayer graphene by shaving one layer at a given time.[29] Structures can be patterned in multilayer graphene. The resolution of the sizes is high. The variety of display configurations and use with less simple devices are more. The graphene material is coated with zinc and then the coating is rinsed using hydrochloric acid. The acid treatment results in removal of the top layer and the underlaying layers are recovered. Zinc can be applied in different patterns and shapes. Stripping of nanosheets is achievable using hydrochloric acid. The possibility of stripping SG came about when scientists used the zinc acid process in order to reduce graphene to graphane by hydrogenation. Peeling a single layer of graphene material in a robust and simple manner can lead to interesting applications. New device exploration as a matter of routine may include local graphene peeling. This method can result in less islands.

Other Metal Substrates

Graphene can be grown on polycrystalline metal substrates. Other than copper, iron, ruthenium, cobalt, rhodium, iridium, nickel, platinum, and gold can be used as metal substrates.

Multilayer graphene can be grown using the CVD method at low pressure and low temperatures of ~600 to 800°C on iron substrates. Low exposure of the surface to acetylene leads to reduced surface coverage and diminished graphene thickness. Thinner graphene is found to form in thicker films, ~6–20 nm of iron.[30] When iron films are less than 5 nm in thickness, the graphene forms separately from the substrate. Graphene thickness was found to increase with increase in growth temperature.[31] However, the number of defects were low.

Graphene can be grown on ruthenium sputter coated on SiO_2.[32] In place of SiO_2, sapphire can be used. Morphology of graphene films deposited on silica was found to be columnar with aligned grains exposing a flat (0001) surface facet. Single-crystalline structure of graphene was found when grown on sapphire. High temperatures of ~800 to 950°C and high vacuum are needed for monolayer formation that is continuous across the grain boundaries. They are slightly "wavy" in appearance. Underlying ruthenium needs to be etched by use of a solution of ceric ammonium nitrate and acetic acid for domain sizes less than 60 μm. The graphene domains are etched downward and then sideways and the graphene is released in order to form a continuous film.

Polycrystalline cobalt can be used as a substrate for growing graphene films.[33] Cobalt deposited on SiO_2/Si c-plane sapphire may also be used.

Monolayer formation of graphene was achieved at high temperatures during the CVD using methane when cobalt deposited on c-plane sapphire was used. But when SiO_2/Si was used, the film thickness was nonhomogeneous and multilayer graphene was formed. One plausible explanation is the high density of cobalt grain boundaries in the SiO_2/Si sample. CVD of acetylene over cobalt has led to nonuniform graphene thickness. The CVD process was conducted at low pressures and moderate to high temperatures of ~800 to 1,000°C. Shorter growth times and increased temperatures are needed for lower graphene thickness. Attributable factors are higher desorption coefficient of carbon at higher temperatures and reduced carbon exposure. Decrease in cobalt substrate thickness was found to decrease the thickness of graphene. Eighty percent of monolayer graphene was found in the case of 100-nm cobalt films. Use of single-crystalline cobalt over sapphire has enabled monolayer graphene formation. Nucleation at grain boundaries can be attributable to formation of multilayer graphene.

Rhodium (111) face is effected using repeated annealing under ultrahigh vacuum.[34] Graphene is then formed on the rhodium foils. When the carbon solubility limit in rhodium is reached, graphene islands can be seen to begin to form. These islands coalesce and as a result complete surface coverage with monolayer graphene is formed. The growth is by carbon segregation when the metal surface is exposed to benzene vapors. Cooling caused multilayer formation. Growth temperature is realized as an important parameter for controlling graphene thickness.

Iridium (111) films deposited on sapphire and yttria-stabilized zirconium oxide (YSZ) can be used as substrates for SG formation.[35] The process used is pulsed laser deposition and electron beam evaporation. Greater than 95% monolayer graphene coverage was achieved at 677°C by CVD of ethylene under ultrahigh vacuum conditions. The carbon exposure time was 10 minutes. Temperature-programmed growth of graphene on YSZ by exposure of acetone was used to confirm that the onset of graphene formation is at ~327°C and the monolayer formation is pronounced upon heating to 727°C.

Graphene can be grown on polycrystalline nickel at atmospheric pressure.[36] Continuous films of graphene that extend across nickel grain boundaries can be formed. Slow annealing and higher deposition temperature have been found to improve graphene quality. A parametric study including parameters such as growth temperature, gas mixing ratio, and growth time on CVD of acetylene/hydrogen was completed. Higher temperature, higher hydrogen concentration, and shorter growth time were found to be causative in the formation of fewer-layer graphene with lower defects. Cooling rate and shorter growth times during the CVD of methane have led to thinner graphene films. Nickel-grown graphene used in electronic devices is enabled by etching of nickel using aqueous $FeCl_3$. Diffusion and segregation of carbon from underlying amorphous carbon or nanodiamond can be used to grow graphene on nickel surfaces.

Platinum surfaces have been used as substrates[37] for the preparation of single-layer and bilayer graphenes. CVD of methane at low pressures and high temperatures of ~1,000 to 1,050°C was used. Higher temperatures resulted in thicker films. Increased growth time/increased flow rate of carbon sources was seen to result in the formation of irregularly shaped islands of bilayer graphene.

Growth of graphene on copper substrates has been discussed in the section "Synthesis in an Annular Plug Flow Reactor and the Roll-to-Roll Transfer Process for Larger Areas." Graphene thickness was found to be independent of copper substrate thickness. Centimeter-scale graphene films have been grown on copper substrates as opposed to micron-scale domains using other substrates.[39] Curving of copper foils in a quartz reactor as discussed in the section "Synthesis in an Annular Plug Flow Reactor and the Roll-to-Roll Transfer Process for Larger Areas" can curtail graphene inhomogeneity. 99.999% monolayer coverage of graphene on copper is seen. In Figure 5.5 a high-resolution image of graphene grown on copper is shown. The hexagonal ring is hard to miss. Temperature had little effect on graphene thickness when a gold substrate was used[40] in the temperature range ~850 to 1,050°C.

Summary

The cost of production of graphene is ~$60 per square inch of copper substrate and needs to be reduced. Graphene can be rolled into thin films by a transfer process. The process comprises adhesion of polymer supports to the graphene on a copper foil, etching of copper layers,

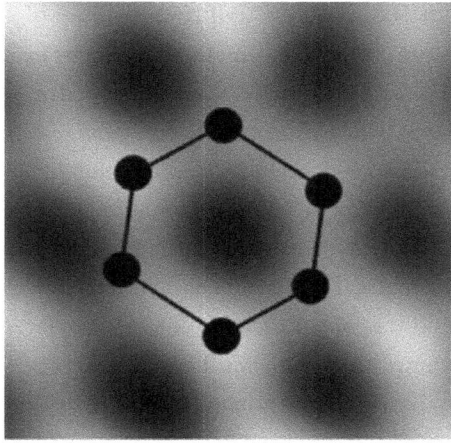

Figure 5.5 A cropped high-resolution image of graphene grown on copper.[38]

and release of graphene layer onto a target substrate. Graphene can be manufactured in an APFR. The outer tube is an 8-inch quartz tube and the inner tube is wrapped with a 7.5-inch copper foil. The operating temperature is 1,000°C, pressure is 90 to 460 mTorr, and cooling rate is 10°C s^{-1}. 150 to 250 mm min^{-1} transfer rates by thermal release are achieved at 90 to 120°C. Scalability of the process is high. Methane dissociates, and the acetylene that is formed is adsorbed onto copper and then carbon is formed by surface reaction of copper carbide. Adsorption kinetics may be assumed to be of the Langmuir–Hinshelwood type. Acetylene decomposition is autocatalytic and is a free-radical process. Axial flow in the annulus is laminar. Diffusion times are high and directed radially inward. The effectiveness factor is given by Eqn. (5.11). Heat effects can be assumed to be due to radiation.

Scientists at UCLA[41] have developed a one-step, solvothermal reduction method for production of GO dispersions in NMP. Hummers' procedure was used for oxidation of graphite and SG was formed by reduction of SGO. SGO is dispersed in DMF. Graphene can be deposited on the carbonizing catalyst from a carbon source. A liquid-carbon-based material can be brought in contact with the carbonizing catalyst and graphene can be implanted in the catalyst. Carburization conditions are: Vigorous stirring at temperatures of 100 to 400°C and contact times of 10 min to 48 h. Samsung[42] used acid solutions to obtain graphene by exfoliation.

Scientists at Harvard University[43] have developed an ion-implantation method for graphene synthesis. A precise dosage of carbon atoms is introduced into polycrystalline nickel films. Carbon solubility in nickel increases with temperature. This can be used for crystallization of carbon atoms on the nickel surface. Fifteen percent less carbon is used and layer control is higher using this process. Diffusion of carbon ions in nickel can be modeled using Stokes–Einstein formulation for ion mobility. Acceleration of ions can be accounted for as shown in Eqn. (5.13).

Graphene can be made from ethanol using a reduction method similar to the Wolff–Kishner process using $NaBH_4$, sodium hydroxide, and SDS surfactant. A scalable high-throughput route to GS was developed at the University of California, Riverside.[44] A hypereutectic carbon and nickel system (Figure 5.2) was formed. Graphene is formed by crystallization. Spin coating of PMMA was used in the transfer and etching steps. Richard E. Smalley Institute for Nanoscale Science and Technology developed a chemical–thermal method for unzipping CNTs into graphene. A suspension of MWCNTs is acid washed and then GO is reduced into graphene. An SDS surfactant was used. Otherwise, potassium was melted over MWCNTs under vacuum at 250°C. Potassium-split MWCNTs were exfoliated using chlorosulfonic acid.

Alternating layers of graphene and graphane with nanoribbon morphology were patented by IBM, Armonk, NY.[45] This morphology is desired in the electronics industry. In situ electrochemical reduction of a GO film on an ITO substrate was shown by Liu et al. (2011)[46] using a CH1660C workstation and a three-electrode system. Flash cooling right after the CVD step was reported by Miyata et al. (2010).[47] Nickel was reacted with ethanol. Cooling occurred from 900°C to 560°C and the cooling time was 10 s. Graphene formed on the nickel surface and then was transferred.

In order to cut the cost of production, graphene platelets are used in place of CNTs. High furnace temperatures can increase the utility costs. An expensive washing step is eliminated. The process in the study by Jang et al.[48] comprises three steps of pre-pressurization, intercalation, and exfoliation. A gas species is intercalated and then graphene is exfoliated in the exfoliation chamber. Fluidization is used. Gas intercalation times range from a few minutes to a few hours. Intercalation by gas and exfoliation by reduction of pressure save cost. Diffusion time of gases in a 100-μm flake can be estimated at 11.5 days. When the flake size is reduced to 1 μm, the diffusion time reduces to 100 s. Refined estimates of diffusion times can be made using damped wave transport and relaxation models. A steady-state concentration can be realized after a said time. For 1-μm flakes, the time to steady state is estimated as 22.2 μs.

Samsung[49] has patented a lower-cost process to make graphene in the form of a shell structure. The shape of the catalyst is used to determine the shape of the shell such as sphere, cylinder, polyhedron, and plate. Coal tar pitch and PAN can be used as starting materials to make graphene.[50]

Fullerenes are the third allotropic form of carbon. Euler stability can be met using some pentagons in order to close a sheet of hexagons. A combustion synthesis method is used in a PFR to make fullerene. The second-generation process is lower in cost. Other methods for fullerene synthesis are counterdiffusion from solution to form needle fibers, supercritical extraction, electric arc process, chemical route by synthesis of corannulene from naphthalene. Acid catalyzed aldol trimerization of ketone can be used to zip oligoarene into a soccer-ball structure.

CNTs are rolled GS about their needle axis. Five methods of synthesis of CNTs are arc discharge, laser ablation, CVD, HIPCO, and surface-mediated growth of vertically aligned tubes.

The electrochemical method for graphene preparation has been developed at the University of Texas, Austin, TX. Graphene flake synthesis was made possible by tuning the parameters that were responsible for swelling of the electrode during battery charging. Electrically conducting paper can be made from graphene ink. Chemists at Rice University have developed an etching method.[51] Metal substrates other than copper can be used such as iron, ruthenium, cobalt, rhodium, and so forth.[52]

CHAPTER 6

Properties

Chapter Objectives

- Magnetic properties
- Surface properties
- Quantum properties
- Electrical properties
- Mechanical properties
- Hexagonal onion rings
- Electrorheological fluids
- Promoter of nickel catalyst

Magnetic Properties

With the high electron mobility, single-layer graphene (SG) can be expected to possess excellent magnetic properties. A current-carrying conductor has a magnetic field associated with it. Edge states, such as "armchair" and "zigzag" configurations of graphene sheets, and adsorbed or intercalated species in the graphene materials can have a salient effect on the magnetic properties. Graphene samples prepared by different methods were tested for magnetic properties. All samples showed divergence between field-cooled (FC) and zero-field-cooled (ZFC) data.

The divergence disappears upon application of 1 T magnetic field. Magnetic hysteresis was seen in all samples. Curie–Weiss behavior was seen in all samples.[1] Negative Weiss temperature and absence of spin-glass behavior was observed. Dominant ferromagnetic interactions coexist with antiferromagnetic interactions in all of the samples. The source of magnetism is not clear in graphene materials. Defects and edge effects can be critical factors. Molecular charge transfer was studied using adsorption studies and the magnetization values were found to decrease with adsorption. Evidence that magnetism found in graphene is intrinsic can be

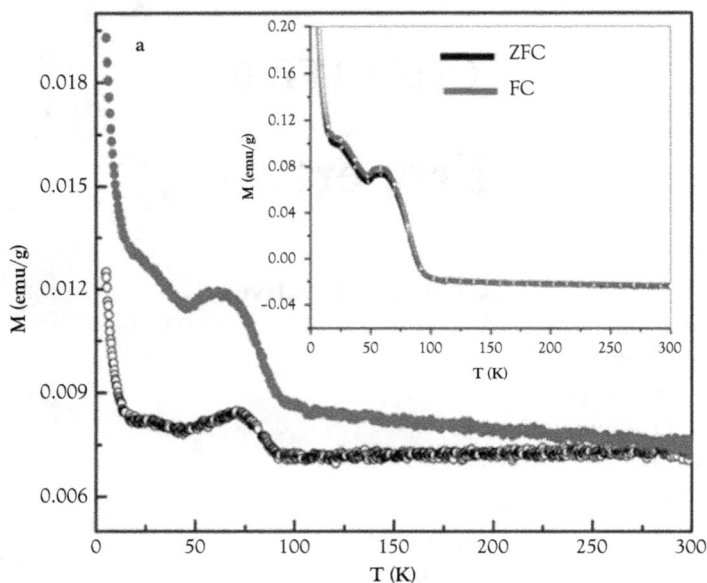

Figure 6.1 Magnetization change with temperature in SG at 500 Oe
Source: Reproduced with permission from Rao, C. N. R., Sood, A. K., Voggu, R., and Subrahmanyam, K. S. (2010). Some novel attributes of graphene. *Journal of Physical Chemistry Letters* 1(2), 572–580.

found in the concentration-dependent effects. Dominant ferromagnetic and antiferromagnetic interactions can be seen to coexist as in phase-separated systems by examination of magnetic properties of graphene.

Anisotropic alignment of ferromagnetic domains may be expected in graphene materials such as graphene nanoribbons (GNRs). Electrical percolation is concentration dependent for isotropic materials. GNRs have high aspect ratios. They are both concentration and alignment dependent. It is estimated that optimal conductivity for a fixed concentration of carbon nanotubes (CNTs) is achievable when they are made to be partially aligned. Intrinsic magnetism in GNRs can be put to good use only in structurally perfect materials such as zigzag-edged GNRs. The presence of edge defects is found to destabilize the magnetism. Stacks of GNRs can be considered similar to CNTs in two dimensions. Similar electrical percolation behavior may be expected. More p orbital delocalization of electrons may be expected in the stacks of GNRs. Genoria et al.[2]

intercalated iron and other ferromagnetic materials between the stacks of GNRs. An anisotropic response to an external magnetic field can be expected. Dispersible ferromagnetic GNRs are needed for alignment in liquids. Edge functionalization can increase the dispersibility without sacrifice of magnetic and electronic conductivities.

Suspensions of iron-intercalated tetradecyl GNR (Fe@TD-GNR) stacks in chlorobenzene drop-cast onto a SiO_2/Si substrate and dried inside a magnetic field are shown in the right-hand-side image of Figure 6.2. The left-hand-side image in Figure 6.2 shows the solution that was dried outside the magnetic field. Iron was intercalated between edge-functionalized GNR stacks in order to make Fe@TD-GNRs. Intercalated iron was imaged using a transmission electron microscope (TEM). The iron concentration was increased using optimal preparation conditions. Thermogravimetric analysis (TGA) and X-ray photoelectron spectroscopy (XPS) studies were used to measure the iron content. The deviation from estimates using XPS and that of TGA can be used to measure the degree of intercalation of iron. Directional alignment of Fe@TD-GNRs in a magnetic field was enabled by anisotropy and magnetization. Alignment was confirmed using scanning electron microscopy (SEM). Enhanced electrical percolation was found for the 5 wt% suspension. Enhanced percolation, solubility, and conductivity can be used to prepare energy-related devices, transparent touch screens, carbon fiber spinning, coating, polymer composites, commercial conductive fluids, lithium-ion batteries, and ultracapacitors.

No magnetic field   Magnetic field

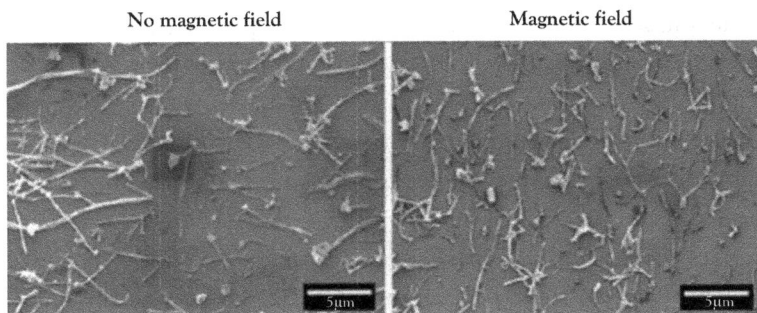

Figure 6.2 SEM images of Mitsui-originated Fe@TD-GNRs

Surface Properties

The theoretical estimate of surface area in SG is ~2,600 m^2 g^{-1}. The measured Brunauer–Emmett Teller (BET)[3] of few-layer graphene was found to be 270 to 1550 m^2 g^{-1}. Hydrogen can be stored in high-surface-area graphenes. High-surface-area graphene structures can be expected to possess high adsorption properties. They can be used for energy storage devices. Adsorption and desorption of gases such as carbon monoxide, nitrogen dioxide, ammonia, and water can be detected using graphene micron-level sensors. The measured surface areas of single-layer and multi-layer graphenes were found to be similar to each other. The Langmuir and BET surface areas for four different samples are reported in the study by Dervishi et al.[4] The samples were prepared using acetylene or methane with different catalyst systems. Surface area of samples prepared using acetylene was found to be higher than those of samples prepared using methane. This may be due to the lower molecular diameter of acetylene compared with that of methane.

Quantum Hall Effect

Graphene can be used to examine quantum phenomena in two dimensions. Consider a current flowing in the plane of a two-dimensional graphene sheet. A magnetic field is applied in the normal direction to the current flowing in the sheet. This induces a transverse voltage that is found to increase in discrete quantum steps. This is called the *quantum hall effect*. This can be seen in metals at low temperatures. This phenomenon occurs in graphenes at ambient temperatures. Two different types of quantum hall effects have been found in bilayer graphene and SG.

Some aspects of quantum electrodynamics (QED) can be tested using graphene sheets. Otherwise, these experiments would need high-energy particle accelerators. Waves of electric charges across the hexagonal lattice of graphene can be considered as quasiparticles. They are analogous to photons—massless and have quantum character of electrons. Special theory of relativity is needed to explain these phenomena. Klein paradox can be seen in graphene.[5] Relativistic charged particles can tunnel through any energy barrier without any obstacle. The energy barrier may be wide or high. As per QED, such particles generate a corresponding antiparticle that possesses opposite charge. In graphenes, the quasiparticles and holes are equivalent to particles and antiparticles. Electrons and positrons in particle physics experiments are similar to this pair.

Electrical Properties

As shown in Figure 6.2, the polycyclic aromatic structure of graphene gives rise to delocalized electrons from the benzene rings in abundance. Increased electron mobility may be expected in SG without any obstacle. Electron mobility within 300 times of that of light in vacuum has been reported. Wallace in 1947 predicted the band structure seen in graphene. The Pi and Pi* states are noninteracting. The Pi states form the valence band and the Pi* states form the conduction band. These two bands touch at six points called Dirac/neutrality points. The honeycomb lattice structure of graphene made of hexagonal carbon rings has two carbon atoms per unit cell of the lattice. By symmetry arguments, the six points are reduced to a pair K and K' independent of each other. Zero bandgap can be inferred from the band structure. Band structure is symmetric about the Dirac point electrons and holes in graphene can be expected to have the same properties. The linear dispersion as that of light can be written as follows (Planck's law);

$$E = chk \tag{6.1}$$

where c is the speed of light, h is the Planck's constant, and k is the wave vector. The relativistic Dirac Hamiltonian can be written as:

$$H = v_F hk\sigma \tag{6.2}$$

where v_F is the Fermi velocity of graphene and σ is the spinor-like wave function. Spinor character does not come from the spin but arises from the two carbon atoms per unit cell of the lattice. The energy of a relativistic particle can be written as follows:

$$E = \sqrt{\left(mc^2\right)^2 + c^2 p^2} \tag{6.3}$$

where m is the rest mass of the particle, p its momentum, and c the velocity. From the linear dispersion of electrons in graphene, they can be considered to have zero rest mass. They can be considered as Dirac Fermions.

Carriers in graphene have been found to move with a Fermi velocity of 1 million m s^{-1} in the ballistic transport regime. Backscattering through long-range interactions such as impurities with charge or phonons (from vibrations of the lattice) is not allowed. Elastic mean free path in neat samples are a few hundred nanometers. Acoustic phonon scattering is considered weak. Optical phonons have a high frequency of ~1,600 cm^{-1}. Scattering from them is relevant at high applied electric fields.[6] When carriers undergo elastic and inelastic collisions in long graphene channels, the ballistic transport phenomenon becomes diffusive. Elastic scattering may be due to Coulomb forces between charged impurities, defects, adsorbates, surface roughness, and ripples on the graphene surface. Phonons of graphene can contribute to inelastic scattering. Thermally excited surface phonons can lead to scattering when combined with graphene carriers. Increase in carrier density generally results in a decrease in carrier mobility. In exfoliated graphene where substrate interactions are eliminated, the carrier mobility was clocked at 200,000 cm^2 V^{-1} s^{-1}. In insulators such as amorphous silica, the mobility is in the order of thousands to tens of thousands square centimeter per volt per second. Graphenes made by other methods such as CVD, epitaxial, have lower mobilities of ~1,000 s cm^2 V^{-1} s^{-1}.

Graphene oxide (GO) can be prepared in an aqueous dispersion form. It can comprise single layer or multilayers with oxygenated functional groups. The functional groups are usually present in the edges and the bottom plane. The sheets can be reassembled into thin films or paper form in the free-standing mode. Stacks of GO sheets are formed by flow-directed assemblage of the aqueous GO dispersion. Property enhancement can be expected by use of a GO dispersion. They can be used to form flexible electrodes for lithium-ion batteries and supercapacitors. Larger GO sheets have been found to have higher electrical conductivity due to lower intersheet contact resistance. A systematic study to better understand the structure–property relationship in GO materials was undertaken.[7] GO size can be controlled using different methods of synthesis such as chemical exfoliation, electrochemical method, oxidation, shaking and sonication, and tweaking the pH value. Ultralarge single-layer GO sheets were prepared with well-aligned, conductive, and strong graphene papers by Lin et al. (2012).[8] Several important characteristics of GO papers were identified such as, degree of oxygenation, C/O ratio, Raman D/G peak intensity ratio, and quality of alignment. Four different sorted GO sheets, namely (1) small GO (S-GO); (2) large GO (L-GO); (3) very large GO (VL-GO); and (4) ultralarge GO (UL-GO), were studied. The failure rate of the samples was described using two-parameter Weibull distribution.[9] The mean and standard deviations of the areas and perimeters of the four different groups differed by an order of magnitude. Change in GO size manifests as different surface chemistry. The degree of oxidation of GO sheets can be deduced from the C/O atomic ratio. Theoretical predictions can be used to confirm that the carboxyl groups are present at the edges and hydroxyl and epoxide groups are present in the bottom plane. A higher perimeter-to-area ratio is seen with the smaller GO sheets. These have been found with more carboxyl groups (edge effect). sp^2 hybridized carbon can be confirmed by the G-band in Raman spectra. D-band can be seen to be activated at the onset of double resonance scattering near the K point of the Brillouin zone on account of defects. Defects in the structure can be studied by use of the peak intensity ratio of D and G bands. As the GO area in the samples was found to increase, the ratio of the D–G peaks were found to decrease. One attributable factor is the expectation of fewer defects in edge boundaries in L-GO.

The 2θ values from X-ray diffraction (XRD) analysis can be used to confirm a rising trend with increase in GO size and a falling trend in d spacing between adjacent sheets. These spacing values were found to be three times greater than the interplanar spacing of planar sheets of carbon in graphite. Interlayer spacing is seen to be a function of degree of oxidation. Stability of the dispersion is maintained at higher zeta potential. The zeta potential can be used to provide a measure of the surface charges present in the particles and their effect on agglomeration. Self-assembly effects were noted during vacuum filtration of GO sheets of different sizes. Quantitative measures of degree of alignments in GO papers were obtained using fast Fourier transform (FFT) analysis. Electrical conductivity of the samples was found to increase with increase in GO sizes as shown in Figure 6.3.

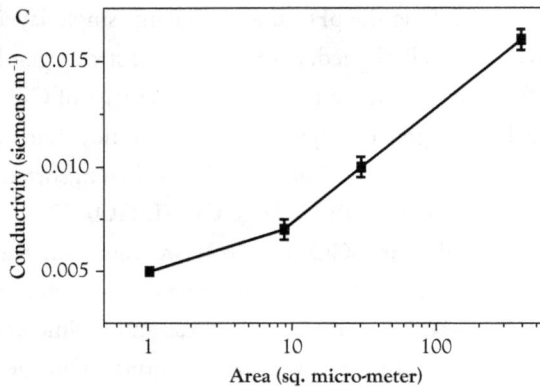

Figure 6.3 Electrical conductivity as a function of GO size

GO papers upon thermal reduction by use of hydrogen iodide (HI) were found to have an increase in electrical conductivity with increase in GO size. In addition to this, the thermal reduction process has resulted in 185 to 283% improvement in the electrical conductivity. This may be because of the higher intersheet contact resistance. The carboxyl groups at the edges have been found to have an insulating effect. Increase in electrical conductivity in the sample has been found with more compactness and better alignment of the GO domains. The measured electrical conductivity values were found to be higher than the theoretical projections. Two synergistic factors may be attributable to the observed increases in electrical conductivity: (1) Ul-GO sizes and (2) effective reduction by high-temperature annealing.

FFT analysis was used to study the effect of degree of alignment of GO papers on the electrical conductivity of the material. Gray values present in an SEM image indicate that pixel by pixel (u,v) is converted into a two-dimensional frequency domain $F (u,v)$. The intensity in the frequency domain is a function of angular dependence on spatial alignment patterns. Higher intensity indicates a higher degree of anisotropy. Similar FFT analyses have been used for studies of collagen bundles and carbon-fiber reinforced composites. The experimental data were found to fit well with the Cauchy–Lorentz distributions. Better alignment was found for larger sizes of GO.

Graphene paper with its excellent conductive properties can be used in portable electronics, energy storage devices, photovoltaic cells, and so on and so forth. Electronic conductivity in graphene can be as low as 200 siemens cm^{-1} depending on the number of defects, grain boundaries, residual oxygen-containing functional groups, and larger intersheet contact resistance. Long one-dimensional silver nanowires are combined with graphene sheets in order to improve their electrical conductivity.[10] Graphene sheets prepared using chemical vapor deposition (CVD) and silver nanowires are filtered without any addition of adhesives/surfactants. Conductivity increases to 3,189 siemens cm^{-1} have been seen. The silver nanowires are prepared by the solvothermal route at 160°C for 2.5 h. The dimensions of the silver nanowires are 150 to 200 nm in diameter and length of about a few microns. XRD peaks were used to confirm the face-centered cubic (fcc) lattice structure for the silver nanowires. Graphene was prepared by both Wurtz-type reductive coupling (WRC) process and by CVD process.

Simple, vacuum filtration of silver nanowires and graphene suspension through a filter membrane was shown by Chen et al.[11] Graphene sheets and silver nanowires are attached to the cellulose fibers by electrostatic interactions. With increases in filtration times, complete coverage of cellulose surfaces with silver nanowires and graphene sheets was found. A sandwich structure was formed from silver nanowire, graphene, and filter membrane. Shaping into desired structures and angles is possible. The filter membrane is peeled off from the composite structure and transferred into another substrate such as polyethylene terephthalate (PET), fluorine-doped tin oxide (FTO). The composite on the transferred PET substrate was bent over 100 cycles evincing high flexibility. The silver content in the composite had a salient effect on the electrical conductivity of the composite.

SEM images were used to confirm the microscale flat morphology found in graphene sheets.

Only a few wrinkles could be found and the sheets were a few atomic layers thick. Oriented layered structure formation was striking. Random, crumpled silver nanowires were found to be dispersed in the composite. Pore filling and bridging in graphene sheets by silver nanowires were found. Composite electrical conductivity was measured as 1,097 siemens cm^{-1},

3.77 times higher than that measured for graphite flakes, 365.7 times higher than that measured for WRC graphene, and 914.2 times higher than that of reduced GO. One attributable factor in higher electrical conductivity in graphene is the crystalline structure formed when grown on a three-dimensional nickel foam template. Less obstacles can contribute to increased traffic of electrons. The electrical conductivity of the composite was found to increase with increase in silver nanowire content of upto 50% as shown in Figure 6.4. Pure silver conductivity is about 630,000 siemens cm^{-1} at 20°C. The asymptotic limit of the composite conductivity at 100% silver content can be seen from Figure 6.4 to be not equal to the pure silver conductivity. The asymptotic limit is significantly less than the pure silver conductivity. The reason for this is not clear. Experiments are needed with pure silver dispersed with small fractions of graphene

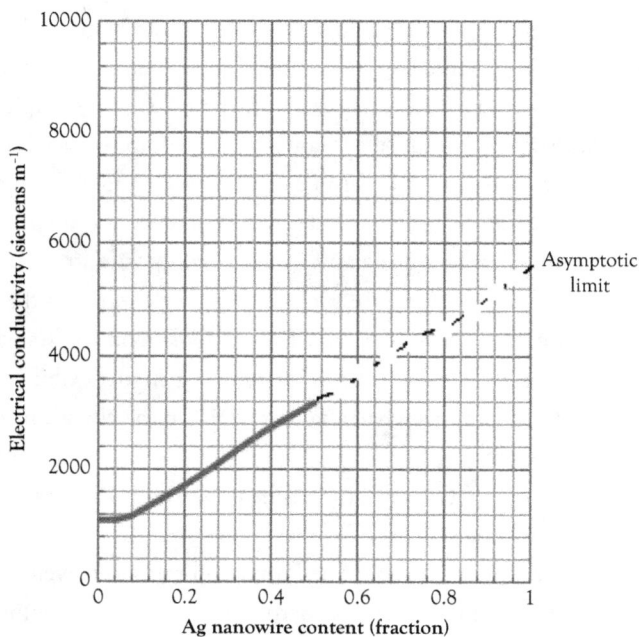

Figure 6.4 Effect of nanowire fraction on a composite's conductivity

sheets and the electrical conductivity is to be measured. The values may be closer to the higher 630,000 siemens cm^{-1} value of pure silver.

Electrorheological Fluids

ER fluids are also called smart fluids and can be used in automatic transmission systems of automobiles. They possess a special property of changing viscosity as a function of electrical field applied. Sometimes, the change in viscosity is by a factor of 10. GO dispersions can be used to make ER fluids. Rapid microstructural transition from a fluid state to a solid-like state within 1000 ns during the application of electrical field can be seen in smart fluids. Other potential application areas include torque transducers, vibration attenuators, control systems, and ER polishing. Rheological properties of GO-based ER fluids were measured using a rotational rheometer and an inductance, capacitance, and resistance (LCR) meter.[12] Rheological flow curves and shear viscosity along with dielectric spectra were reported. Particle chain structure formation can be linked to the fluid's viscoelastic property. GO is fabricated from graphite powder using strong oxidizing agents by a modified Hummers' method. The oxidation process is confirmed by the color change. The GO slurry is dispersed in distilled water and exfoliated by sonication using an ultrasonic generator. GO sheets were separated by centrifugation and washed and dried in the oven. GO particles were dispersed in silicone oil by sonication. The fibrillation phenomenon was observed using optical microscopy. In Figure 6.5, the transparent morphology of GO can be seen.

In Figure 6.6, the ER effect can be seen. When the electric field is applied, chain-like structures are formed and the material behaves in a solid-like fashion. Without the electric field, the GO appears dispersed in Figure 6.6(a). Both the storage modulus and loss modulus were found to increase with an increase in the electrical field over a broad frequency range. This can be due to increase in interparticle forces. The ER fluid made out of dispersed GO was found to exhibit solid-like behavior when the storage and loss modulus trends became a plateau with increase in frequency. This viscoelastic nature of the solid may be due to the particle chain structure formation as shown in Figure 6.6(b).

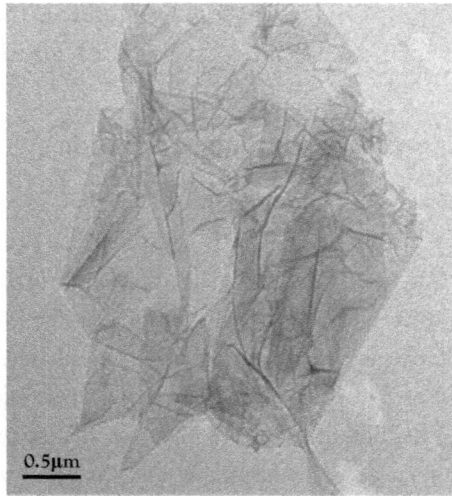

Figure 6.5 A TEM image of GO

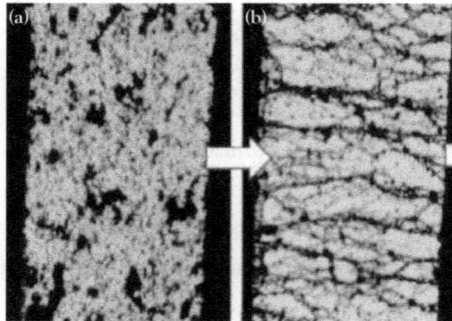

Figure 6.6 ER effect—optical microscope images (a) without and (b) with the electric field

Properties of Hexagonal Graphene Onion Rings

The electronic and other interesting properties of graphene depend on the spatial structures. The synthesis method of CVD of large-scale, single-layer, two-dimensional graphene on copper substrates was discussed in Chapter 5. There are some limitations to the seed-induced two-dimensional growth mechanism of graphene on copper. In situ complex spatial structures are not possible to be formed. Hybrid structures such as graphene–CNTs, CNTs, CNT–graphene ribbons, graphene–graphane, and hexagonal boron nitride (h-BN)–graphene structures have been reported. Building three-dimensional graphene structures on the edges of two-dimensional graphene by control of partial pressure of hydrogen was reported by Yan et al.[13] Hexagonal graphene onion rings (HGoRs) can be made in this manner.

Top-side view of a HGoR is shown in Figure 6.7. GNRs can be synthesized from HGoRs by removal of the top layer in a graphene domain using Ar plasma. The important bandgap that is attributable as the elusional attribute of graphene from being directly used as field-effect transistor (FET) can be introduced by shaping graphene into GNRs. HGoRs are grown on copper foils that are cleaned by electrochemical polishing and high-pressure annealing. The polished copper is loaded into the vacuum chamber and annealed for 30 min at 1,074°C. The flow rate of hydrogen was 500 sccm and that of methane was 7 sccm. The vacuum pressure was 500 Torr. The reaction was allowed for 35 min. After the reaction, the copper foils were quickly removed from the hot zone of the furnace by use of a magnetic rod and cooled to room temperature.

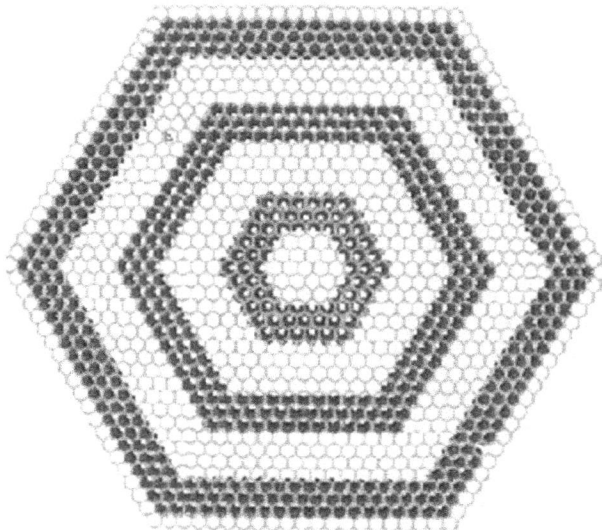

Figure 6.7 Topology of a HGoR

Samples were characterized using Raman spectroscopy, surface electron microscopy (SEM), atomic force microscopy (AFM), and TEM.

The HGoRs can be transferred to SiO_2 substrates. In Figure 6.7, the concentric growth of graphene rings can be seen. The blue-colored hexagons lie below the light-colored hexagons. The mechanism of onion ring formation proposed is as follows: A monolayer graphene domain forms at first on the surface of a copper foil. Another graphene layer begins to nucleate and grow on the edges of the monolayer graphene. This leads to the formation of one-dimensional hexagonal graphene ribbon rings along the edge of the monolayer. A new edge was exposed due to the rapid growth of the monolayer graphene domain. Repeated growth leads to the formation of HGoR. A three-dimensional structure is also formed. Nucleation, growth, and coalescence are the important items in the mechanism.

HGoRs can be used to make GNRs, which can be used as FETs. SG is not suitable for lithium ion battery storage. This is because of the weak binding of lithium on the graphene surface. HGoRs can be more suitable for lithium-ion battery storage. This is because the lithium binding energy between ribbon layers is 1.8 eV and 3.3 eV at the ribbon edges. These values are higher than the cohesive energy of lithium, which is ~1.6 eV.

Promoter for Nickel-Based Catalysts

Nickel-based catalysts are used in hydrogenation and dehydrogenation of hydrocarbons, steam reforming, and partial oxidation of methane. Nickel is used as an alloy constituent in high-temperature steels and constructional materials for tubular reactors in steam crackers and steam reformers. The lower price and good performance properties make nickel an attractive choice for a catalyst material. During operation, coking or carbon deposition can lead to the deactivation of the catalyst. Coking is also found in the reactor walls that contain nickel. Surface carbon has been found to be an intermediate in many important catalytic reactions such as methanation. Surface diffusion of carbon has been found. Carbon chemisorption on nickel has been studied from first-principles[14] density functional theory. Graphene overlayer formation on a nickel catalyst during steam reforming has been predicted using the model. The catalytic mechanism of Fischer–Tropsch synthesis using carbon on Miller plane

(111) of nickel. Attempts were made to better understand the coking mechanism. The following steps have been identified: (1) on-surface carbon formation from adsorbed hydrocarbon molecules; (2) formation of nickel carbide by carbon atoms diffusing into bulk; and (3) formation of graphitic carbon. Some studies have shown immediate formation of graphene layers from adsorbed carbon. A VASP simulation package was used[15] in order to calculate the chemisorption energies using periodic spin-polarized density functional theory. The nickel catalyst was modeled as four-layered slabs separated by 12-A° vacuum layers. The top two layers are allowed to relax. A $7 \times 7 \times 1$ grid of k points for single monolayer coverage was used in the Brillouin-zone integrations. The grid was changed to a $5 \times 5 \times 1$ grid for 0.25 ML. On-surface carbon, bulk carbon, and a graphene overlayer were studied. On-surface adsorption sites considered were fcc, hexagonal close packed (hcp), bridge, and atop. It was found that carbon binding energies increase strongly with decreasing coverages. Octahedral sites were found to be preferred over the tetrahedral sites. Four adsorption models were considered: (1) carbon atoms at hcp and fcc sites; (2) carbon atoms atop and in hcp sites; (3) carbon atoms atop and in fcc sites; and (4) carbon atoms atop. The binding energy of chemisorbed graphene was -760 kJ mol^{-1} and was found to be more stable. Lattice constant of graphene was found to change slightly upon adsorption from 2.46 to 2.49 A° and matches the lattice constant of a nickel substrate. Four different coking mechanisms were evaluated. Carbon diffusion into the bulk is thermodynamically favorable. From the studies, it was concluded that bulk carbon was preferred over on-surface adsorbed carbon. The build-up of bulk carbon can be expected to have an effect on the catalytic properties of nickel and lead to catalyst deactivation. An overlayer of graphene on a (111) nickel plane was found to be more stable than the bulk carbon. Formation of a graphene overlayer was found to obey higher-order kinetics. Coking resistance of Ni can be increased by use of these calculations.

Mechanical Properties

Elastic properties and breaking strength of a SG membrane were measured by nanoindentation in an AFM.[16] AFM is a special case of scanning probe microscopy (SPM). SPM evolved as a part of a revolution in the characterization and analysis of materials at the nanoscale. Neither light nor electrons is used to obtain the image. Topographical images are obtained at the atomic scale. SPMs are different from optical and electron microscopes. A specimen is characterized and the surface features are obtained. More than 1 billion times magnification is achievable. SPM techniques can be used in order to study nanoscale features. AFM was invented in 1986. The specimen surface is scanned using a microscale cantilever with a probe with a sharp tip at its end. Radius of curvature of the tip is of the order of magnitude of a few nanometers. Deflection of the cantilever is caused by the force between the tip and the sample. Deflection caused may be described using Hooke's law of elasticity. Molecular forces such as van der Waals, hydrogen bonding, surface tension, electrostatic, and magnetic can be measured using the AFM. The specimen surface is excited using a laser source. The laser light is reflected by the cantilever and captured using photodiodes (Figure 9.6 in Sharma[17]). Deflection of the cantilever is measured using a piezoresistive material that is made up of a Wheatstone bridge circuit. Feedback control is used in order to control the tip-to-sample distance. Tapping and imaging are the two modes of operation. The probe tip is raster scanned using a piezoelectric scanner.

The concept of break strength of a brittle material was introduced by Griffith in 1921. He established the relationship between the change of potential energy of a brittle system and crack growth and free energy of a created surface. Defects and flaws within the material determine the break strength of the material. One of Griffith's observations[18] was that a fiber with molecules in "single file" would possess the theoretical maximum tensile strength. Maximum stress that can be supported by the material prior to failure in a material without defects is called the intrinsic strength. Griffith measured experimentally the breaking strength of a series of glass fibers with different diameters and extrapolated an intrinsic strength of about $E/9$ to an atomic radius. E is the elastic stiffness of the material under uniaxial tension. Lee et al.[19] used AFM nanoindentation in order to measure the mechanical properties of SG membranes suspended over open holes. Sample geometry can be defined precisely; a sample less sensitive to single defect and sheet is clamped around the hole circumference. A 5×5 mm array of circular holes with diameter 1 to 1.5 μm and depth of 500 nm was patterned on a Si substrate with an epilayer of 300-nm thickness made of SiO_2. Methods used for patterning are nanoimprint lithography and reactive ion etching.

A "space elevator" is an exotic structure than can be realized if macroscopic fibers close to the theoretical strength can be realized. Graphite flakes were mechanically deposited on the substrate. SG was confirmed by optical microscopy and thickness of graphene was confirmed using Raman spectroscopy. The AFM used was XE-100, from Park Systems. The material of construction used for a cantilever was diamond. Tip radii of cantilevers used were 16.5 nm and 27.4 nm as measured after and before indentation. TEM was used to measure the deflection. Cantilever spring constants were calibrated using a reference cantilever. No hysteresis was found from the tests. The center of each film was indented. Force-displacement measurements were found to be reproducible. The elastic modulus of the material was obtained for samples with different graphite flakes, membrane diameters, displacement rates, and indenter tip radii. They were within the statistical rigor independent of the specimen/test parameters.

Elastic response of the graphene can be expected to be nonlinear. Maxima or the point where the curve reaches a maximum value in the stress–strain response can be used to define the intrinsic breaking stress. Elastic response can be used to deduce the existence of an energy potential that is a function of strain. This function can be expressed as a Taylor power series of strains. Linear elastic response is given by the lowest-order quadratic term. Nonlinear elastic behavior is given by the third term. Isotropic elastic response under uniaxial extension for graphene may be expressed as follows:

$$\sigma = E\varepsilon + D\varepsilon^2 \tag{6.4}$$

where ε is the uniaxial Lagrangian strain and σ is the symmetric second Piola–Kirchhoff stress and D is the third-order elastic modulus. The value of D is usually negative. This would mean that at high tensile strain, the stiffness decreases and at high compressive strain, the stiffness increases. Components of a second-order, fourth-rank stiffness tensor was used to determine E. D was determined from both the second-order, fourth-rank stiffness tensor and third-order, sixth-rank stiffness tensor. Nonlinear elastic response was confirmed as appropriate using numerical simulations of graphene sheets and nanotubes. Salient features of the elastic behavior of graphene were captured using this thermodynamically rigorous nonlinear form of the stress–strain response. Intrinsic stress is obtained from the maximum of the elastic stress–strain response. Its functional form is as follows:

$$\sigma_{int} = -\frac{E^2}{4D}$$
$$\varepsilon_{int} = -\frac{E}{2D}$$

(6.5)

E and D are obtained from the experimental data. E is obtained from the force-displacement data and D is inferred from the experimental breaking force.

As graphene is a two-dimensional material, the strain energy density is normalized by the area of the graphene sheet in place of the volume. Two-dimensional stress, σ^{2D}, and elastic constants E^{2D} and D^{2D} possess units of Nm^{-1}. The force displacement behavior is approximated by the model:

$$F = \sigma_0^{2D}(\pi a)\left(\frac{\delta}{a}\right) + E^{2D}\left(q^3 a\right)\left(\frac{\delta}{a}\right)^3$$

(6.6)

where F is the applied force, δ is the deflection at the center point, σ_0^{2D} is the pretension in the film, v is Poisson ratio, and

$$q = \frac{1}{\left(1.05 - 0.15v - 0.16v^2\right)}$$

(6.7)

They measured an in-plane Young's modulus of elasticity of 1.0 ± 0.1 TPa and third-order elastic stiffness $D = -2.0$ TPa. The corresponding $\sigma^{2D} = 42$ N m^{-1} and $D^{2D} = -690$ N m^{-1}. The graphene thickness assumed was 3.35 A°. A suspended graphene film was found to have a Young's modulus of elasticity value of 0.5 TPa. Elastic deformation and failure strength of graphene were simulated and a nonlinear response was predicted for strains above 10%.

Summary

SG can be expected to have excellent magnetic properties on account of the delocalized electrons. Divergence between FC and ZFC data is shown in Figure 6.1. Curie–Weiss behavior was observed. Dominant ferromagnetic and anti-ferromagnetic interactions can be found to coexist. Anisotropic alignment of ferromagnetic domains can be expected in GNRs. Genoria et al.[20] intercalated iron between the stacks of GNRs. Figure 6.2 shows the dried samples of Fe@TD-GNRs with and without the magnetic field. Alignment was confirmed using SEM. Applications can be found in touchscreens, carbon fiber spinning, coating, polymer composites, commercial conductive fluids, ultracapacitors, and lithium-ion batteries.

The measured BET surface area of graphene was 270 to 1550 m^2 g^{-1} versus theoretical estimate of ~2,600 m^2 g^{-1}. Graphenes can be used in storage of hydrogen. They can be used as sensors during the adsorption and desorption processes. Surface areas of single- and multilayer graphenes were found to be similar to each other.

A quantum hall effect can be seen in graphenes at room temperature. Experiments that would need high-energy particle accelerators can be tested using graphene sheets. Waves of electric charges across the hexagonal lattice of graphene can be considered as quasiparticles. Klein paradox can be seen in graphene.

Delocalized electrons in the polycyclic aromatic structure of graphene result in increased electron mobility without any obstacle. Electron mobility within 300 times of that of light in vacuum has been reported. Wallace in 1947 predicted the Pi and Pi* states and band structure in graphene. Linear dispersion of light is given by Planck's law. Relativistic Dirac Hamiltonian can be written in terms of Fermi velocity of graphene and spinor-like wave function. Carriers in graphene have been found to move with a Fermi velocity of 1 million m s^{-1} in the ballistic transport regime. Ballistic to diffusion transition can be expected upon collisions. Electron mobility was found to vary between graphene samples made from different methods. Higher mobilities are found in exfoliated samples and lower mobilities have been found in samples from CVDs and epitaxial methods. GO can be dispersed in water. Electrical conductivity as a function of GO size is shown in Figure 6.3. FFT analysis of SEM images was used to study the effect of degree of alignment of GO papers on the electrical conductivity of the material. Experimental data were found to fit well with the Cauchy-Lorentz distributions.

Composites of graphene and silver were prepared in order to increase the conductivity. The trend in the experimental data for electrical conductivity versus silver nanowire fraction appears to be linear. When extrapolated to 100% silver nanowire, the electrical conductivity of the composite was ~5,600 siemens cm^{-1}. This is lower than the pure silver conductivity of 630,000 siemens cm^{-1}.

GO aqueous suspensions can be used to make ER fluids. Solidification in 1000 ns from fluid state to solid state can be seen upon application of electrical field. Rheological properties of a GO-based ER fluid were studied. The ER effect is shown in Figure 6.2 under an optical microscope.

HGoRs are built on the edges of two-dimensional graphene by control of partial pressure of hydrogen (Figure 6.7). HGoRs are grown on copper foils. A graphene overlayer on a nickel catalyst is used for studying the coking mechanism using a VASP simulation package. Graphene has been used as a promoter along with the nickel catalyst.

Nanoindentation was used to obtain the elastic properties and breaking strength of SG. The operating principle of AFM is discussed. Griffith observed that a fiber with molecules in "single file" would possess the theoretical maximum tensile strength. Elastic response of the graphene can be expected to be nonlinear. D, a third-order elastic modulus, is introduced in Eqn. (6.4). Force displacement behavior of two-dimensional graphene is described using a cubic equation in Eqn. (6.6). An in-plane Young's modulus of elasticity of 1.0 ± 0.1 TPa and third-order elastic stiffness $D = -2.0$ TPa were measured.

About the Author

Dr. Kal Renganathan Sharma, P. E., received his BTech in chemical engineering from Indian Institute of Technology, Chennai, India, in 1985 and MS and PhD degrees in chemical engineering from West Virginia University, Morgantown, WV, in 1987 and 1990, respectively. His postdoctoral research training was under former chair and Prof. R. Shankar Subramanian, Clarkson University, Potsdam, NY, in chemical engineering. He is the author of 13 books, 548 conference papers, 46 journal articles, 7 book chapters, 3 review articles, and 113 other presentations. He has instructed 2,830 students in India and the United States in 104 courses. He is a Fellow of the Indian Chemical Society and is listed in Who's Who in America. He has reviewed over 30 journal articles. He serves on editorial boards of four journals: (1) *ChemXpress*, Rajkot, India; (2) *Caspian Journal of Applied Science and Research*, Penang, Malaysia; (3) *Journal of Scientific Research and Reports*, Science Domain International, Wilmington, DE; and (4) *Journal of Pharmaceutical and Biomedical Sciences*, Delhi, India. He serves as an adjunct professor at Houston Community College-Central, Lone Star College-North Harris, and Texas Southern University, Houston, TX. He has held high-level positions in academia. He has received cash awards from Monsanto Plastics Technology, Indian Orchard, MA; SASTRA University, Thanjavur, India; and Prairie View A & M University, Prairie View, TX. He has received 14 press citations. His works have been cited more than 240 times in refereed journals. One of his papers has been downloaded more than 753 times. His books have been cataloged in more than 3,500 libraries world over. He has introduced four dimensionless groups and one statistical distribution into the literature and has served as a cochair at nine conferences.

His first journal article with J. W. Zondlo, E. A. Mintz, P. Kneisl, and A. H. Stiller, "Preparation of an Ultra-Low Ash Coal Extract under Mild Conditions," *Fuel Processing Technology*, Vol. 18, 3, 1988, 273–278, has been cited 80 times. This is an article on fixed carbon. He has introduced four dimensionless groups in the literature: frequency number, storage modulus, *Sharma number* (mass), and momentum number. His study on acceleration effects in motion of free electrons provided a *new perspective* to the propagation of heat and how the thermal conductivity changes with temperature. One of the citations of his book had this to say in their article "Sharma *has proved* that Taitel paradox is because of poor use of initial conditions . . ." Nanoscale effects of heat conduction in time were studied. Computer simulation studies leading to useful correlations were presented. A paper presented in an American Chemical Society conference in 1997 at Las Vegas, NV, on small rubber particle morphology in ABS thermoplastics has been cited in *Encyclopedia of Chemical Technology, Encyclopedia of Polymer Science & Technology*, 2003/2004. *Bioinformatics: Sequence Alignment and Markov Models,* McGraw Hill Professional, New York, NY, 2009, has been cataloged in 530 libraries according to the Worldcat.org. This book has been cited in creation Wikipedia under bioinformatics in English and Wikipedia in Portuguese under Proteina and MAFFT. *Nanostructuring Operations in Nanoscale Science and Engineering* has been cataloged in 502 libraries. This book has been listed as required text in the syllabus of Biomedical Engineering at the University of Mumbai, India.

Notes

Chapter 1

1. Geim (2010).
2. Zondlo (2013).
3. Askeland, Fulay, and Wright (2011).
4. Kroto (1996).
5. Ijima (1991), pp. 56–58.
6. Rode, Gamaly, and Luther-Davies (1997), pp. 135–144.
7. Renganathan, Zondlo, Mintz, Kneisl, and Stiller (1988), pp. 273–278.
8. Sharma (2013).
9. Markoff (2011).
10. Chang (2007).
11. Chang (2007).
12. Naik (2013).
13. Geim (2009), pp. 1530–1534.

Chapter 2

1. Sharma (2010).
2. Flynn (2011).
3. Podila, Rao, Tsuchikawa, Ishigami, and Rao (2012).
4. Sundarajan et al. (2008).
5. Petkewich (2008).
6. Scipioni, Stern, Notte, Sijbrandij, and Griffin (2008).
7. Lemme et al. (2009).
8. Xu, Chen, Wang, Tang, Li, and Hsiao (2011).
9. Dubeck (2005).
10. Hernandez et al. (2008).
11. Zhang and Vauthey (2007).
12. Chen et al. (2013).
13. Duan, Daniels, Niu, Sahi, Hamilton, and Romano (2008).
14. Mirkin, Piner, and Hong (2004).
15. Naitoh et al. (2011).
16. Blick (2000).
17. Liu, Auilar, Hao, Ruoff, and Armitage (2011).
18. Aita, Yakovlev, Cayton, Mirhoseini, and Aita (2005).

19. Liebler and Kim (1993).
20. Rao, Sood, Voggu, and Subrahmanyam (2010).
21. Zhang, Chen, Yuan, Ji, Cheng, and Qiu (2013).
22. Raghavan (2004).
23. Kishore et al. (2013).
24. Sharma (2010).
25. Xu, Chen, Wang, Tang, Li, and Hsiao (2011).
26. Zhang and Vauthey (2007).

Chapter 3

1. Sharma (2010).
2. http://physicsworld.com/cws/article/news/2009/jan/05/graphene-transistor-speeds-up
3. http://physicsworld.com/cws/article/news/2009/jan/05/graphene-transistor-speeds-up
4. http://physicsworld.com/cws/article/news/2010/feb/05/graphene-transistor-breaks-new- record
5. Dery et al. (2012).
6. Meyer, Chuvilin, Algara-Siller, Biskupek, and Kaiser (2009).
7. Ruoff (2008).
8. Luo, Jang, and Huang (2013).
9. Luo, Jang, and Huang (2013).
10. Sharma (2007).
11. Cohen-Tanugi and Grossman (2012).
12. Li, Zou, Pan, and Sun (2010).
13. Hwang et al. (2012).
14. Hwang et al. (2012).
15. Kim and Huang (2012).
16. http://www.eetasia.com/ART_8800687153_480200_NT_101c0d89.HTM?click_from=8800103309,9950188816,2013-07-09,EEOL,ARTICLE_ALERT
17. Choi, Lahiri, Seelaboyina, and Kang (2010).
18. Sharma (in press).
19. Choi, Lahiri, Seelaboyina, and Kang (2010).
20. Schmidt (2012).
21. Bian et al. (2013).
22. Qi et al. (2011).
23. Sharma (2010).
24. Stankovich et al. (2006).
25. Sturzel, Kempe, Thomann, Mark, Enders, and Mulhaupt (2012).

26. Sturzel, Kempe, Thomann, Mark, Enders, and Mulhaupt (2012).
27. Wilson (2010).
28. Sharma (2005).
29. Sharma (2009).
30. Gurney, Marinero, and Pisana (2012).
31. Paul and Sharma (2011).
32. Steenackers et al. (2011).
33. Kucsko et al. (2013).
34. Sharma (2013a).
35. Sharma (2013b).
36. Sawyer (2012).
37. Howlader et al. (2012).
38. Craig (2005).
39. Sharma (2012a).
40. Sharma (2010).
41. Shishodia, Sethi, and Aggarwal (2005).
42. Frietas (1999).
43. Ishiyama, Sendoh, and Arai (2002).
44. Mathieu, Martel, Yahia, Soulez, and Beaudoin (2003).
45. Klocke (2010).
46. Regan, Zettl, and Aloni (2011).
47. Dimitrov, Peng, Xue, and Wang (2009).
48. Seifert, Samuelson, Ohlsson, and Borgstrom (2011).
49. Sharma (2012b).
50. Sharma (2013a).
51. Sharma (2013b).
52. Sharma (2013a).
53. Sharma (2013a).
54. Sharma (2013a).
55. Sharma (2013a).
56. Sharma (2013a).
57. Sharma (2013a).
58. Sharma (2013a).
59. Sharma (2013a).

Chapter 4

1. Geim (2010).
2. Suzuki (2011).
3. Kroto (1997).
4. Sharma (2012).

5. Zhang, Wu, Li, and Yang (2011).
6. Zhang, Wu, Li, and Yang (2011).
7. Zhang, Wu, Li, and Yang (2011).
8. Booth et al. (2008).
9. Booth et al. (2008).
10. Booth et al. (2008).
11. Booth et al. (2008).
12. Booth et al. (2008).
13 Askeland, Fulay, and Wright (2011).
14. Girit et al. (2009).
15. Girit et al. (2009).
16. Nair (2012).
17. Girit et al. (2009).
18. Bonilla and Carpio (2011).
19. Sharma (2005).
20. Bonilla and Carpio (2011).
21. Sharma (2005).
22. Zhang and Liu (2011).
23. Booth et al. (2008).
24. Girit et al. (2009).
25. Girit et al. (2009).
26. Bonilla and Carpio (2011).
27. Sharma (2005).
28. Bonilla and Carpio (2011).
29. Sharma (2005).

Chapter 5

1. Bae et al. (2010).
2. Fogler (2006).
3. Sharma (2007).
4. Byrd, Stewart, and Lightfoot (1961).
5. Rao, Subrahmanyam, Matte, and Govindaraj (2011).
6. Dubin (2010).
7. Park et al. (2009).
8. Shin, Choi, and Yoon (2012).
9. Garaj, Hubbard, and Golovchenko (2010).
10. Sharma (2005).
11. Cussler (1997).
12. Sharma (2005).
13. Bard and Faulkner (1980).

14. Sharma (2005).
15. Singh, Iyer, and Giri (2011).
16. Amini, Garay, Liu, Balandin, and Abbaschian (2010).
17. Jiao, Zhang, Wang, Diankov, and Dai (2009).
18. Kosynkin, Lu, Sinitskii, Pera, Sun, and Tour (2011).
19. Cohen, Dimitrakopoulos, Grill, and Winieff (2013).
20. Liu, Ou, Wang, Liu, and Yang, (2011).
21. Miyata, Kamon, Ohashi, Kitaura, Yoshimura, and Shinohara (2010).
22. Jang, Zhamu, and Guo (2010).
23. Sharma (2005).
24. Sharma (2005).
25. Sharma (2007).
26. Choi, Shin, and Yoon (2011).
27. Jang and Huang (2006).
28. Sharma (2010).
29. Dimiev, Kosynkin, Sinitskii, Slesarev, Sun, and Tour (2011).
30. Edwards and Coleman (2013).
31. An, Lee, and Jung (2011).
32. Sutter, Albrecht, Camino, and Sutter (2009).
33. Wang et al. (2010).
34. Rut'kov, Kuz'michev, and Gall (2011).
35. Mueller et al. (2011).
36. Edwards and Coleman (2013).
37. Edwards and Coleman (2013).
38. Edwards and Coleman (2013).
39. Edwards and Coleman (2013).
40. Edwards and Coleman (2013).
41. Dubin (2010).
42. Shin, Choi, and Yoon (2012).
43. Garaj, Hubbard, and Golovchenko (2010).
44. Amini, Garay, Liu, Balandin, and Abbaschian (2010).
45. Cohen, Dimitrakopoulos, Grill, and Winieff (2013).
46. Liu, Ou, Wang, Liu, and Yang (2011).
47. Miyata, Kamon, Ohashi, Kitaura, Yoshimura, and Shinohara (2010).
48. Jang, Zhamu, and Guo (2010).
49. Choi, Shin, and Yoon (2011).
50. Jang and Huang (2006).
51. Dimiev, Kosynkin, Sinitskii, Slesarev, Sun, and Tour (2011).
52. Edwards and Coleman (2013); An, Lee, and Jung (2011); Sutter, Albrecht, Camino, and Sutter (2009); Wang et al. (2010); Rut'kov, Kuz'michev, and Gall (2011); Mueller et al. (2011).

Chapter 6

1. Rao, Sood, Voggu, and Subrahmanyam (2010).
2. Genoria, Peng, Lu, Hoelscher, Novoselov, and Tour (2012).
3. Rao, Subrahmanyam, Matte, and Govindaraj (2011).
4. Dervishi et al. (2012).
5. Panesor.
6. Avouris (2010).
7. Lin et al. (2012).
8. Lin et al. (2012).
9. Weibull and Stockholm (1951).
10. Chen et al. (2013).
11. Chen et al. (2013).
12. Zhang, Liu, Choi, and Kim (2012).
13. Yan et al. (2013).
14. Xu and Saeys (2007).
15. Xu and Saeys (2007),
16. Lee, Wei, Kyasr, and Hone (2008).
17. Sharma (2010).
18. Griffith (1921).
19. Xu and Saeys (2007).
20. Genoria, Peng, Lu, Hoelscher, Novoselov, and Tour (2012).

References

Aita, C. R., Yakovlev, V. V., Cayton, M. M., Mirhoseini, M., & Aita, C. R., Yakovlev, V. V., Cayton, M. M., Mirhoseini, M and Aita M. (2005). *Self repairing ceramic coatings, US Patent 6,869,701*. Shorewood, WI.

Amini, S., Garay, J., Liu, G., Balandin, A. A., & Abbaschian, R. (2010). Growth of large-area graphene films from metal-carbon melts. *Journal of Applied Physics, 108*, 094321.

An, H., Lee, W. J., & Jung, J. (2011). Graphene synthesis on Fe foil using thermal CVD. *Current Applied Physics, 11*(4), S81–S85.

Askeland, D. R., Fulay, P. P., & Wright, W. J. (2011). *The science and engineering of materials* (6th ed.). Stamford, CT: Cengage Learning.

Avouris, P. (2010). Graphene: Electronic and photonic properties and devices. *Nano Letters, 10*, 4285–4294.

Bae et al. (2010). Roll-to-roll production of 30 inch graphene films for transparent electrodes. *Nature Nanotechnology, 5*, 574–578.

Bard, J., & Faulkner, R. (1980). *Electrochemical methods: Fundamentals and applications*. Hoboken, NJ: John Wiley & Sons.

Bian et al. (2013). Covalently patterned graphene surfaces by a force-accelerated Diels–Alder reaction. *Journal of the American Chemical Society, 135*(25), 9240–9243. doi:10.1021/ja4042077

Blick, R. H. (2000). Microwave spectroscopy on quantum dots. In H. S. Nalwa (Ed.), *Handbook of nanostructured materials and nanotechnology* (Vol. 2, pp. 309–343). Amsterdam, Netherlands: Academic Press, Elsevier Science, Spectroscopy and Theory.

Bonilla, L. L., & Carpio, A. (2011). Theory of defect dynamics in graphene: Defect groupings and their stability. *Continuum Mechanics and Thermodynamics, 23*(4), 337–346.

Booth et al. (2008). Macroscopic graphene membranes and their extraordinary stiffness. *Nano Letters, 8*(8), 2442–2448.

Byrd, B. R., Stewart, W., & Lightfoot, E. (1961). *Transport phenomena*. Hoboken, NJ: John Wiley & Sons.

Chang, K. (2007, April 10). Thin carbon is in: Graphene steals nanotubes' allure. *New York Times*.

Chen et al. (2013). Highly conductive and flexible paper of 1D silver-nanowire-doped graphene. *ACS Applied Materials & Interfaces, 5*(4), 1408–1413.

Choi, J. H., Shin, H. J., & Yoon, S. M. (2011). *Process of preparing graphene shell, US Patent 8,075,950*. KR: Samsung Electronics Co.

Choi, W., Lahiri, I., Seelaboyina, R., & Kang, Y. S. (2010). Synthesis of graphene and its applications: A review. *Critical Reviews in Solid State and Materials Sciences*, *35*(1), 52–71.

Cohen, G., Dimitrakopoulos, C. D., Grill, A., & Winieff, R. L. (2013). *Graphene nanoribbons, method of fabrication and their use in electronic devices, US Patent 8,361,853*. Armonk, NY: International Business Machines Corp.

Cohen-Tanugi, D., & Grossman, J. C. (2012). Water desalination across nanoporous graphene. *NANO Letters*, *12*, 3602–3608.

Craig, J. J. (2005). *Introduction to robotics: Mechanics and control* (3rd ed.). Upper Saddle River, NJ: Pearson Prentice Hall.

Cussler, E. L. (1997). *Mass transfer*. Cambridge, UK: Cambridge University Press.

Dervishi et al. (2012). Few-layer nano-graphene structures with large surface areas synthesized on a multifunctional Fe:Mo:MgO catalyst system. *Journal of Materials Science*, *47*(4), 1910–1919.

Dery et al. (2012). Nanospintronics based on magnetologic gates. *IEEE Transaction on Electron Devices*, *59*(1), 259–262.

Dimiev, A., Kosynkin, D. V., Sinitskii, A., Slesarev, A., Sun, Z., & Tour, J. M. (2011). Layer-by-layer removal of graphene for device patterning. *Science*, *331*(6021), 1168–1172. doi:10.1126/science.1199183

Dimitrov, D. V., Peng, X., Xue, S. S., & Wang, D. (2009). *Spin oscillatory device, US Patent 7,589, 600 B2*. Dublin, IE: Seagate Technology, LLC.

Duan, X., Daniels, R. H., Niu, C., Sahi, V., Hamilton, J. M., & Romano, L. T. (2008). *Methods of positioning and/or orienting nanostructures, US Patent 7,422,980*. Palo Alto, CA: Nanosys Inc.

Dubeck, P. (2005). Nanostructure as seen by the SAXS. *Vacuum*, *80*, 92–97.

Dubin et al. (2010). A one-step, solvothermal reduction method for producing reduced graphene oxide dispersions in organic solvents. *ACS Nano*, *4*(7), 3845–3852.

Edwards, R. S., & Coleman, K. S. (2013). Graphene film growth on polycrystalline metals. *Accounts of Chemical Research*, *46*(1), 23–30.

Flynn, G. M. (2011). Perspective: The dawning of age of graphene. *Journal of Chemical Physics*, *135*(5), 050901-1–050901-7.

Fogler, S. (2006). *Elements of chemical reaction engineering*. Upper Saddle River, NJ: Pearson Prentice Hall.

Frietas, R. A. Jr. (1999). *Nanomedicine, Vol I: Basic capabilities*. Georgetown, TX: Landes Bioscience.

Garaj, S., Hubbard, W., & Golovchenko, J. A. (2010). Graphene synthesis by ion implantation. *Applied Physics Letters*, *97*, 183103.

Geim, A. K. (2010). *Random walk to graphene*. Retrieved December 8, 2010, from Nobel Lecture: http://nobel.se

Geim, K. (2009). Graphene: Status and prospects. *Science*, *324*, 1530–1534.

Genoria, B., Peng, Z., Lu, W., Hoelscher, B. K. P., Novoselov, B., & Tour, J. M. (2012). Synthesis of dispersible ferromagnetic graphene nanoribbon stacks with enhanced percolation properties in a magnetic field. *ACS Nano, 6*(11), 10396–10404.

Girit et al. (2009). Graphene at the edge: Stability and dynamics. *Science, 323,* 1705–1708.

Griffith, A. A. (1921). The phenomena of rupture and flow in solids. *Philosphical transactions of the Royal Society of London. Series A, containing papers of a mathematical or physical character, 221,* 163–198.

Gurney, B. A., Marinero, E. E., & Pisana, S. (2012). *Magnetic field sensor with graphene sense layer and ferromagnetic biasing layer below the sense layer, US Patent 8,189,302.* Amsterdam, The Netherlands: Hitachi Global Storage Technologies.

Hernandez et al. (2008). High-yield production of graphene by liquid phase exfoliation of graphite. *Nature Nanotechnology, 3,* 563.

Howlader et al. (Eds). (2012). *SEER Cancer Statistics Review, 1975–2010.* Bethesda, MD: National Cancer Institute.

http://physicsworld.com/cws/article/news/2009/jan/05/graphene-transistor-speeds-up

http://physicsworld.com/cws/article/news/2010/feb/05/graphene-transistor-breaks-new- record

http://www.eetasia.com/ART_8800687153_480200_NT_101c0d89.HTM?click_from=8800103309,9950188816,2013-07-09,EEOL,ARTICLE_ALERT

Hwang et al. (2012). *Graphene light-emitting device and method of manufacturing the same, US Patent 2012/0068152.* KR: Samsung LED Co.

Ijima, S. (1991). Helical microtubules of graphitic carbon. *Nature, 354,* 56–58.

Ishiyama, K., Sendoh, M., & Arai, K. I. (2002). Magnetic micromachines for medical applications. *Journal of Magnetism and Magnetic Materials, 242–245*(1), 1163–1165.

Jang, B. Z., Zhamu, A., & Guo, J. (2010). *Mass production of nano-scaled platelets and products, US Patent 7,785,492.* Dayton, OH: Nanotek Instruments.

Jang, Z., & Huang, W. C. (2006). *Nano-scaled graphene plates, US Patent 7,071,258.* Dayton, OH: Nanotek Instruments.

Jiao, L., Zhang, L., Wang, X., Diankov, G., & Dai, H. (2009). Narrow graphene nanoribbons from carbon nanotubes. *Nature, 458,* 877–880.

Kim, N. P., & Huang, J. P. (2012). *Graphene nanoplatelet metal matrix, US Patent 8,263,843.* Chicago, IL: Boeing Company.

Kishore et al. (2013). Combustion synthesis of graphene and ultracapacitor performance. *Bulletin of Materials Science, 36*(4), 667–672.

Klocke, V. (2010). *Nanrobot module, automation and exchange, US Patent 2010/0140473A1*, Stamford, CT.

Kosynkin, D. V., Lu, W., Sinitskii, A., Pera, G., Sun, Z., & Tour, J. M. (2011). Highly conductive graphene nanoribbons by longitudinal splitting of carbon nanotubes using potassium vapor. *ACS Nano, 5*(2), 968–974.

Kroto, H. (1997). New horizons in the structure and properties of layered materials. In M. A. Serio, D. M. Gruen, & R. Malhotra (Eds), *Synthesis and characterization of advanced materials*. Washington, DC: ACS Symposium Series, American Chemical Society.

Kroto, H. W. (1996). *Symmetry, space, stars and C_{60}*. Retrieved December 1996, from Nobel Lecture: http://nobel.se

Kucsko et al. (2013). Nanometre-scale thermometry in a living cell. *Nature, 500*, 54–58.

Lee, C., Wei, X., Kyasr, J. W., & Hone, J. (2008). Measurement of the elastic properties and intrinsic strength of monolayer graphene. *Science, 321*, 385–388.

Lemme et al. (2009). Etching graphene devices with helium ion beam. *ACS Nano, 3*(9), 2264–2676.

Li, H., Zou, L., Pan, L., & Sun, Z. (2010). Novel graphene-like electrodes for capacitive deionization. *Environmental Science & Technology, 44*(22), 8692–8697.

Liebler, C. M., & Kim, Y. (1993). *Machining oxide thin-films with an atomic force microscope: Pattern and object formation on the nanometer scale, US Patent 5,252,835*. Cambridge, MA: Harvard College.

Lin et al. (2012). Fabrication of highly-aligned, conductive, and strong graphene papers using ultralarge graphene oxide sheets. *ACS Nano, 6*(12), 10708–10719.

Liu, S., Ou, J., Wang, J., Liu, X., & Yang, S. (2011). A simple two-step electrochemical synthesis of graphene sheets film on the ITO electrode as supercapacitors. *Journal of Applied Electrochemistry, 41*, 881–884.

Liu, W., Auilar, R. V., Hao, Y., Ruoff, R. S., & Armitage, N. P. (2011). Broadband microwave and time-domain terahertz spectroscopy of chemical vapor deposition grown graphene. *Journal of Applied Physics, 110*, 083510 (01–05).

Luo, J., Jang, H. D., & Huang, J. (2013). Effect of sheet morphology on the scalability of graphene-based ultracapacitors. *ACS NANO, 7*(2), 1464–1471.

Markoff, J. (2011, June 10). IBM lab created high-speed circuits made more cheaply. *New York Times*.

Mathieu, J. B., Martel, S., Yahia, L., Soulez, G., & Beaudoin, G. (2003, September). *MRI systems as a means of propulsion for a microdevice in blood vessels*. Proceedings of the 25th Annual International Conference of the IEEE, Engineering in Medicine and Biology, Cancun, Mexico, 3419–3422, Vol. 4. doi:10.1109/IEMBS.2003.1280880

Meyer, J. C., Chuvilin, A., Algara-Siller, G., Biskupek, J., & Kaiser, U. (2009). Selective sputtering and atomic resolution imaging of atomically thin boron nitride membranes. *Nano Letters, 9*(7), 2683–2689. doi:10.1021/nl9011497

Mirkin, C. A., Piner, R., & Hong, S. (2004). *Methods utilizing scanning probe microscope tips and products therefore produced thereby, US Patent 6,827,979.* Evanston, IL: Northwestern University.

Miyata, Y., Kamon, K., Ohashi, K., Kitaura, R., Yoshimura, M., & Shinohara, H. (2010). A simple alcohol-chemical vapor deposition synthesis of single layer graphenes using flash cooling. *Applied Physics Letters, 96,* 263105.

Mueller et al. (2011). Epitaxial growth of graphene on Ir(111) by liquid precursor method. *Physics Review B, 84,* 075471.

Naik, G. (2013, August 24). Wonder material ignites scientific gold rush; atom-thin graphene beats steel, silicon; a patent land rush. *Wall Street Journal.*

Nair et al. (2012). Spin-half paramagnetism in graphene induced by point defects. *Nature Physics, 8,* 199–202.

Naitoh et al. (2011). STM observation of graphene formation using sic-on-insulator substrates. *Surface Review and Letters, 18*(5), 163–167.

Panesor, T. (n.d.). *Graphene: A new form of carbon with scientific impact and technological promise.* London, UK. Retrieved from www.iop.org

Park et al. (2009). Colloidal suspensions of highly reduced graphene oxide in a wide variety of organic solvents. *Nano Letters, 9,* 1593–1597.

Paul, W., & Sharma, C. P. (2011). Blood compatibility and biomedical applications of graphene. *Trends in Biomaterials & Artificial Organs, 25*(3), 91–94.

Petkewich, R. (2008). Say hello to helium ion microscopy, chemical and engineering news. *Science and Technology, 86*(47), 38–39.

Podila, R., Rao, R., Tsuchikawa, R., Ishigami, M., & Rao, A. M. (2012). Raman spectroscopy of folded and scrolled graphene. *ACS Nano, 6*(7), 5784–5790.

Qi et al. (2011). Enhanced electrical conductivity in polystyrene nanocomposites—Low graphene content. *ACS Applied Materials and Interfaces, 3*(8), 3130–3133.

Raghavan, V. (2004). *Materials science and engineering: A first course.* New Delhi, India: Prentice Hall of India.

Rao, C. N. R., Sood, A. K., Voggu, R., & Subrahmanyam, K. S. (2010). Some novel attributes of graphene. *Journal of Physical Chemistry Letters, 1*(2), 572–580.

Rao, C. N. R., Subrahmanyam, K. S., Matte, H. S. S. R., & Govindaraj, A. (2011). Graphene: Synthesis, functionalization and properties. *Modern Physics Letters, 25*(7), 427–451.

Regan, B. C., Zettl, A. K., & Aloni, S. (2011). *Nanocrystal powered nanomotor, US Patent 7,863, 798 B2,* Oakland, CA.

Renganathan, K., Zondlo, J. W., Mintz, E. A., Kneisl, P., & Stiller, A. H. (1988). Preparation of an ultra-low ash coal extract under mild conditions. *Fuel Processing Technology, 18*(3), 273–278.

Rode, V., Gamaly, E. G., & Luther-Davies, B. (1997). Formation of cluster-assembled carbon nano-foam by high-repetition-rate laser ablation. *Applied Physics A: Materials Science & Processing, 70*(2), 135–144. doi:10.1007/s003390050025

Ruoff, R. (2008). New carbon-based material for energy storage. *Power Engineering*, 16.

Rut'kov, E. V., Kuz'michev, A. V., & Gall, N. R. (2011). Carbon interaction with rhodium surface: Adsorption, dissolution, segregation, growth of graphene layers. *Physics of Solid State, 53*(5), 1092–1098.

Sathe, C., Zou, X., Leburton, J-P., & Schulten, K. (2011). Computational investigation of DNA detection using graphene nanopores. *ACS Nano*, 5(11), 8842–8851. doi:10.1021/nn202989w

Sawyer, D. (2012, December 20). *ABC World News*.

Schmidt, C. (2012). The bionic material. *Nature, 483*, S37.

Scipioni, L., Stern, L. A., Notte, J., Sijbrandij, S., & Griffin, B. (2008). Helium ion microscope. *Advanced Materials & Processes, 166*(6), 27–30.

Seifert, W., Samuelson, L. I., Ohlsson, B. J., & Borgstrom, L. M. (2011). *Directionally controlled growth of nanowhishers, US Patent 7,911, 035 B2*, Lund, SE.

Sharma, K. R. (2005). *Damped wave transport and relaxation*. Amsterdam, The Netherlands: Elsevier.

Sharma, K. R. (2007). *Principles of mass transfer*. New Delhi, India: Prentice Hall of India.

Sharma, K. R. (2009). *Bioinformatics: Sequence alignment and Markov models*. New York, NY: McGraw Hill Professional.

Sharma, K. R. (2010). *Nanostructuring operations in nanoscale science and engineering*. New York, NY: McGraw Hill Professional.

Sharma, K. R. (2012). *Polymer thermodynamics: Blends, copolymers and reversible polymerization*. Boca Raton, FL: CRC Press.

Sharma, K. R. (2012a). Nanostructuring of nanorobots for use in nanomedicine. *International Journal of Engineering & Technology, 2*(2), 116–134. Retrieved from http://ietjournals.org/archive/2012/feb_vol_2_no_2/6686111325866989.pdf

Sharma, K. R. (2012b). *Nanorobot drug delivery system for curcumin with increased bioavailability during the treatment of Alzheimer's disease*. Baton Rouge, LA: 68th Southwest Regional Meeting of the American Chemical Society (October/November).

Sharma, K. R. (2013, October 7–9). *On the use of graphene in solar cells.* International Conference and Exhibition on Lasers, Optics and Photonics, San Antonio, TX.

Sharma, K. R. (2013a). On photodynamic therapy of Alzheimer's disease using intrathecal nanorobot drug delivery of *Curcuma longa* for enhanced bioavailability. *Journal of Scientific Research and Reports, 2*(1), 206–227.

Sharma, K. R. (2013b). Nanorobot drug delivery system for curcumin for enhanced bioavailability during treatment of Alzheimer's disease. *Journal of Encapsulation and Adsorption Sciences, 3*(1), 24–34.

Sharma, K. R. (in press).*Comprehensive guide to nanocoatings technology.* Happauge, NY: Nova Science.

Shin, H. J., Choi, J., & Yoon, S. (2012). *Method for exfoliating carbonization catalyst from graphene sheet, method for transferring graphene sheet from which carbonization catalyst is exfoliated to device, graphene sheet and device using the graphene sheet, US Patent 8,133,466.* KR: Samsung Electronics Co.

Shishodia, S., Sethi, G., & Aggarwal, B. G. (2005). Curcumin: Getting back to the roots. *Annals of the New York Academy Sciences, 1056,* 206–217.

Singh, D. K., Iyer, P. K., & Giri, P. K. (2011). Improved chemical synthesis of graphene using a safer solvothermal route. *International Journal of Nanoscience, 10*(1,2), 39–42.

Stankovich et al. (2006). Graphene-based composite materials. *Nature, 44,* 282–286.

Steenackers et al. (2011). Polymer brushes on graphene. *Journal of American Chemical Society, 133,* 10490–10498.

Stolyarova et al. (2007). High-resolution scanning tunneling microscopy imaging of mesoscopic graphene sheets on an insulating surface. *Proceedings of the National Academy of Sciences of the United States of America, 104*(29), 9209–9212. With permission from American Chemical Society.

Sturzel, M., Kempe, F., Thomann, Y., Mark, S., Enders, M., & Mulhaupt R. (2012). Novel graphene UHMWPE nanocomposites prepared by polymerization filling using single-site catalysts supported on functionalized graphene nanosheet dispersion. *Macromolecules, 45*(17), 6878–6887.

Sundarajan et al. (2008). *Microfluidic apparatus, Raman spectroscopy systems, and methods for performing molecular reactions, US Patent No. 7,442,339.* Santa Clara, CA: Intel Corp.

Sutter, E., Albrecht, P., Camino, F. E., & Sutter, P. (2009). Graphene growth on polycrystalline Ru thin films. *Applied Physics Letters, 95,* 133109.

Suzuki, T. (2011). Peierls transition of graphene nanoribbons: Crossover from polyacetylenes to graphene. *Physics of Semiconductor, 1399,* 813–814.

Wang et al. (2010). Synthesis of graphene on a polycrystalline Co film by radio-frequency plasma-enhanced chemical vapor deposition. *Journal of Physics D: Applied Physics*, *43*(45), 455402.

Weibull, W., & Stockholm, S. (1951). A statistical distribution function of wide applicability. *Journal of Applied Mechanics*, *18*, 293–297.

Wilson, E. K. (2010, March). Superconductor is simply organic. *Chemical & Engineering News*, *88*(10), 7.

Xu, J. Z., Chen, C., Wang, Y., Tang, H., Li, Z. M., & Hsiao, B. S. (2011). Graphene nanosheets and shear flow induced crystallization in isotactic polypropylene nanocomposites. *Macromolecules*, *44*(8), 2808–2818.

Xu, J., & Saeys, M. (2007). Coking mechanism and promoter design for Ni-based catalysts: A first principles study. *International Journal of Nanoscience*, *6*(2), 131–135.

Yan et al. (2013). Hexagonal graphene onion rings. *Journal of the American Chemical Society*, *135*(29), 10755–10762. doi:10.1021/ja403915m

Zhang, J., Chen, P., Yuan, B., Ji, W., Cheng, Z., & Qiu, X. (2013). Real-space identification of intermolecular bonding with atomic force microscopy. *Science*, *342*(6158), 611–614. doi:10.1126/science.1242603

Zhang, S., & Vauthey, S. (2007). *Surfactant peptide nanostructures and uses thereof, US Patent 7,179,784.* Cambrige, MA: Massachusetts Institute of Technology.

Zhang, W. L., Liu, Y. D., Choi, H. J., & Kim, S. G. (2012). Electrorheology of graphene oxide. *ACS Applied Materials & Interfaces*, *4*, 2267–2272.

Zhang, W., Wu, P., Li, Z., & Yang, J. (2011). First-principles thermodynamics of graphene growth on Cu surfaces. *The Journal of Physical Chemistry*, *115*, 17782–17787.

Zhang, Y., & Liu, F. (2011). Maximum asymmetry in strain induced mechanical instability of graphene: Compression versus tension. *Applied Physics Letters*, *99*, 241908.

Zondlo, J. W. (2013). Graphite: Structure, properties, and applications. In P. Mukhopadhyay & R. K. Gupta (Eds.), *Graphite, graphene and their polymer nanocomposites*. Boca Raton, FL: CRC Press.

Index

THIS BOOK IS IN OUR NANOMATERIALS COLLECTION

Momentum Press is dedicated to developing collections of complementary titles within specific engineering disciplines and across topics of interest. Each collection is led by a collection editor or editors who actively chart the strategic direction of the collection, assist authors in focusing the work in a concise and applied direction, and help deliver immediately actionable concepts for advanced engineering students for course reading and reference.

Some of our collections include:

- *Manufacturing and Processes*—Wayne Hung, Editor
- *Manufacturing Design*
- *Engineering Management*—Carl Chang, Editor
- *Electrical Power*
- *Communications and Signal Processing*—Orlando Baiocchi, Editor
- *Electronic Circuits and Semiconductor Devices*—Ashok Goel, Editor
- *Integrated Circuit Design*
- *Sensors, Control Systems and Signal Processing*
- *Antennas, Waveguides and Propagation*
- *Thermal Engineering*—Derek Dunn-Rankin, Editor
- *Fluid Mechanics*—Dr. George D. Catalano, Editor

- *Environmental Engineering*—Francis Hopcroft, Editor
- *Geotechnical Engineering*—Dr. Hiroshan Hettiarachchi, Editor
- *Transportation Engineering*
- *Sustainable Systems Engineering*—Dr. Mohammad Noori, Editor
- *Structural Engineering*
- *Chemical Reaction Engineering*
- *Chemical Plant & Process Design*
- *Thermal and Kinetics Topics in Chemical Engineering*
- *Petroleum Engineering*
- *Materials Characterization and Analysis*—Dr. Richard Brundle, Editor
- *Mechanics & Properties of Materials*
- *Computational Materials Science*
- *Biomaterials*

Momentum Press is actively seeking collection editors and authors. For more information about becoming an MP author or collection editor, please visit **http://www.momentumpress.net/ contact** and let us hear from you.

Announcing Digital Content Crafted by Librarians

Momentum Press offers digital content as authoritative treatments of advanced engineering topics, by leaders in their fields. Hosted on ebrary, MP provides practitioners, researchers, faculty and students in engineering, science and industry with innovative electronic content in sensors and controls engineering, advanced energy engineering, manufacturing, and materials science. **Momentum Press offers library-friendly terms:**

- perpetual access for a one-time fee
- no subscriptions or access fees required
- unlimited concurrent usage permitted
- downloadable PDFs provided
- free MARC records included
- free trials

The **Momentum Press** digital library is very affordable, with no obligation to buy in future years.

For more information, please visit **www.momentumpress.net/library** or to set up a trial in the US, please contact **mpsales@globalepress.com**.